I0064881

Agricultural Food Economics: Global Challenges and Developments

Agricultural Food Economics: Global Challenges and Developments

Edited by **Salvador Flores**

SYRAWOOD
PUBLISHING HOUSE

New York

Published by Syrawood Publishing House,
750 Third Avenue, 9th Floor,
New York, NY 10017, USA
www.syrawoodpublishinghouse.com

Agricultural Food Economics: Global Challenges and Developments
Edited by Salvador Flores

© 2016 Syrawood Publishing House

International Standard Book Number: 978-1-68286-036-6 (Hardback)

This book contains information obtained from authentic and highly regarded sources. Copyright for all individual chapters remain with the respective authors as indicated. All chapters are published with permission under the Creative Commons Attribution License or equivalent. A wide variety of references are listed. Permission and sources are indicated; for detailed attributions, please refer to the permissions page and list of contributors. Reasonable efforts have been made to publish reliable data and information, but the authors, editors and publisher cannot assume any responsibility for the validity of all materials or the consequences of their use.

The publisher's policy is to use permanent paper from mills that operate a sustainable forestry policy. Furthermore, the publisher ensures that the text paper and cover boards used have met acceptable environmental accreditation standards.

Trademark Notice: Registered trademark of products or corporate names are used only for explanation and identification without intent to infringe.

Printed in the United States of America.

Contents

Preface   VII

Chapter 1   **Capital accumulation with and without land market liberalization: beyond the "Win-Win" situation**   **1**
George Vachadze

Chapter 2   **Valuation of traits of indigenous sheep using hedonic pricing in Central Ethiopia**   **17**
Zelalem G Terfa, Aynalem Haile, Derek Baker and Girma T Kassie

Chapter 3   **Institutional reforms and agricultural policy process: lessons from Democratic Republic of Congo**   **30**
Catherine Ragasa, Suresh C Babu and John Ulimwengu

Chapter 4   **Performance and profit sensitivity to risk: a practical evaluation of the agro-industrial projects developed by Israeli companies for the CIS and Eastern European countries**   **51**
Gregory Yom Din

Chapter 5   **Bioenergy chain building: a collective action perspective**   **74**
Luigi Cembalo, Francesco Caracciolo, Giuseppina Migliore, Alessia Lombardi and Giorgio Schifani

Chapter 6   **Crop substitution behavior among food crop farmers in Ghana: an efficient adaptation to climate change or costly stagnation in traditional agricultural production system?**   **87**
Zakaria A Issahaku and Keshav L Maharjan

Chapter 7   **Farm technology adoption in Kenya: a simultaneous estimation of inorganic fertilizer and improved maize variety adoption decisions**   **101**
Maurice J Ogada, Germano Mwabu and Diana Muchai

Chapter 8   **Evaluating willingness to become a food education volunteer among urban residents in Japan: toward a participatory food policy**   **119**
Yasuo Ohe, Shinichi Kurihara and Shinpei Shimoura

Chapter 9   **The perfect storm of business venturing? The case of entomology-based venture creation**   **138**
Stefano Pascucci, Domenico Dentoni and Dimitrios Mitsopoulos

Chapter 10   **The adoption of technologies, management practices, and production systems in U.S. milk production**   149
Jeffrey Gillespie, Richard Nehring and Isaac Sitienei

Chapter 11   **Hawai'i's food consumption and supply sources: benchmark estimates and measurement issues**   172
Matthew K Loke and PingSun Leung

Permissions

List of Contributors

Preface

This book has been an outcome of determined endeavour from a group of educationists in the field. The primary objective was to involve a broad spectrum of professionals from diverse cultural background involved in the field for developing new researches. The book not only targets students but also scholars pursuing higher research for further enhancement of the theoretical and practical applications of the subject.

Agricultural food economics is an interdisciplinary field of study which aims to apply the principles and concepts of economics for analysing agricultural productivity and food supply. Agri-food market analysis, livestock management, agri-food policy and trade, agricultural supply chain management, consumer behavior, etc. are some of the diverse topics covered in this book that address the varied branches which fall under this subject. For all readers who are interested in agricultural food economics, the researches and examples included in this book will serve as an excellent guide to develop a comprehensive understanding of the current progress in this field.

It was an honour to edit such a profound book and also a challenging task to compile and examine all the relevant data for accuracy and originality. I wish to acknowledge the efforts of the contributors for submitting such brilliant and diverse chapters in the field and for endlessly working for the completion of the book. Last, but not the least, I thank my family for being a constant source of support in all my research endeavours.

Editor

Capital accumulation with and without land market liberalization: beyond the "Win-Win" situation

George Vachadze

Correspondence:
george.vachadze@csi.cuny.edu
Department of Political Science,
Economics, and Philosophy, College
of Staten Island and Graduate
Center, City University of New York,
2800 Victory Blvd, Staten Island, NY
10314, USA

Abstract

This paper examines the effect of land market liberalization on the dynamics of capital accumulation. It is shown that the land market liberalization, which is accompanied with the transfer of agricultural technology, may not always offer a "win-win" outcome for developed and developing countries. Improved agricultural productivity generates a growth enhancing externality. However, land market liberalization affects the balance between the equalizing force of the diminishing returns technology and the un-equalizing force of the low income elasticity of the agricultural commodity demand. As a result, land market liberalization accompanied with the transfer of agricultural productivity, may not always guarantee a "win-win" outcome for developed and developing countries. If improvement of agricultural productivity is insignificant then land market liberalization can cause "win-lose" situation for developed and developing countries. This result suggests that one should be very careful in a policy proposal designed to foster the process of development through foreign land ownership. It is important to recognize that apart from benefits, foreign land ownership also creates a disadvantage for capital accumulation and causes the magnification of the world income inequality.

JEL classification: F43, O11, R14

Keywords: Dynamics of capital accumulation, Land market liberalization, Transfer of agricultural technology

Background

Food prices, which almost doubled between 2006 and 2008, had a major impact on the perception of food insecurity. Rising food prices hit hard to the balance of payments of many food importing countries. In order to insure themselves, those countries started to acquire farmland under their control. Reports from the International Food Policy Research Institute (IFPRI) estimate that during the last decade tens of millions of acres of farmland has been acquired by countries where the consumption of agricultural commodities far exceeds the production (see Von Braun and Meinzen-Dick (16) for more details). Food-importing countries with land and water constraints but rich in oil resources, such as the Gulf States, or countries with large populations and food security concerns, such as China, Japan, India, and South Korea, are seeking opportunities to produce food and biofuel crops overseas (see Borras et al. (3) for more details).

Land acquisitions are occurring mainly in developing countries like Brazil, Cambodia, Indonesia, Laos, Madagascar, Pakistan, Philippines, Uganda, Sudan and others, where production costs are relatively low and where land and water resources are more abundant than in the investor nation (see Vidal (15), Cotula et al. (5), and Allen (1), for more details).

Many powerful international institutions, such as The World Bank, the European Bank for Reconstruction and Development, and the US Millennium Challenge Corporation, are actively advising national governments in developing countries to allow a large-scale acquisition of domestic land by foreigner investors (see Li (11) and World Bank (17) among others). These international institutions argue that a large-scale land acquisition is a way to reduce poverty through transfer of technology and technical know-how from developed to developing countries and improve the overall efficiency in the agricultural sector in general. While the robust empirical evidence about the impact of foreign land ownership on economic growth is lacking, many developing countries have already simplified their land ownership laws for foreign firms and governments.[a] The speed of land acquisition has provoked opposition from farmers' organizations, human rights groups, and other social movements as they raise concerns that foreign land ownership can have a negative impact on poor local people, who risk losing access to and control over the land on which they depend too much.[b] In addition, unrestricted foreign land ownership could cause a spike in property prices so that citizens of the low and middle class are never able to afford land again.[c]

In the present paper, I take a small step toward reconciling two conflicting views about land market liberalization: one is a "win-win" view, typically encouraged by the World Bank and the UN Food and Agriculture Organization, and the other "win-lose" (win for rich and lose for poor) view, usually supported by numerous non-profit organizations such as La Via Campesina, the Oakland Institute, GRAIN, Food First, and many others (See for example Nierenberg and Pollack (13)). To this end, I analyze a general equilibrium model within which I can investigate the effects of land market liberalization on the dynamics of capital accumulation and on the well-being of a nation. The world economy consists of identical countries that differ in their levels of agricultural sector productivity and the initial capital stock. The model is set up in such a way that, in the absence of land market liberalization, a country with a relatively more productive agricultural sector converges monotonically to a unique steady state and achieves higher capital stock, higher wage income, higher consumption, and higher welfare in general.[d] This result can be tempting to conclude that poor countries will benefit from the land market liberalization if such liberalization comes with the transfer of more efficient agricultural technology. In this paper I demonstrate that, as a result of land market liberalization, the unique steady state can lose its stability property and two asymptotically stable steady states come in existence. The unique steady state can become unstable because, with the integrated land markets, the land price in different countries must move together. This creates a disadvantage for relatively poor countries because of low income elasticity of agricultural demand. The land market will crowd out domestic investment in capital and will create a negative spillover for next generations. That is to say, a low aggregate investment in capital lowers the income of the next generations of the same country, creating a downward spiral of low-income/low-investment in capital. The opposite force will operate in a relatively rich country which because of land market liberalization will face relatively low land prices.

This will create an upward spiral of high-income/high-investment in capital. As a result, a "win-win" outcome is not necessarily guaranteed through the land market liberalization even if such liberalization comes with the improvement of agricultural sector productivity in the relatively poor country.

The paper is organized as follows. Section The model introduces the model with a simple, analytically tractable structure, which is capable of capturing the effect of the land market integration on the dynamics of capital accumulation. Section The autarky case demonstrates the existence and uniqueness of an interior steady state and show its stability property under rational expectations dynamics. Section Two country model considers the world economy in which foreign land ownership is allowed and the productivity of agricultural sector is equalized. In such a set-up, I show that the foreign land ownership right can cause the long-run divergence of otherwise symmetric economies. Section Conclusions summarizes results and concludes.

Methods

The model

Let us consider an infinite-horizon, a standard neoclassical overlapping generations model, modified only to include a land market.[e] The world economy consists of two countries: Country 1 and Country 2. Countries are identical in all respect except for their levels of productivity in the agricultural sector and the initial capital stock.[f] In each period there are two generations alive: young and old. Population size of each generation is constant and normalized to unity. Agents supply one unit of labor while young and consume while old. A single manufacturing commodity is produced by an infinitely lived neoclassical firm which combines two factors of production, capital and labor.

The produced final commodity is usable for consumption and for investment. Per capita output in country $i = 1, 2$, in period t, is $y_{it} = f(k_{it})$, where k_{it} denotes the capital per worker in Country i and $f : \mathbb{R}_+ \to \mathbb{R}_+$ is the production function in intensive form. We assume that the production function satisfies the standard neoclassical properties: f is twice continuously differentiable on $\mathbb{R}_{++}$; continuous, strictly increasing, and strictly concave, on $\mathbb{R}_+$; and satisfies the following boundary conditions, $f(0) = 0$, $\lim_{k \downarrow 0} f'(k) = \infty$, and $\lim_{k \uparrow \infty} f'(k) = 0$. Factor markets are competitive and factor rewards are determined according to the marginal product rule, i.e., at time t, the wage rate and the rate of return on one unit of capital are given by $W(k) := f(k) - kf'(k)$ and $f'(k)$ respectively. Existing capital depreciates fully within a period.

After receiving the wage income $W(k_{it})$, young agents decide how much to invest in capital and how much land to purchase in the competitive land market. Aggregate supply of land in each country is fixed and normalized to unity. Productivity of land $A_i > 0$ in each country is constant over time and is treated as an exogenous parameter. Purchase of x_{it} units of land implies the consumption of $A_i x_{it}$ units of agricultural commodity. After harvesting the agricultural commodity, land ownership can be sold in order to increase the consumption of the manufacturing commodity.[g]

Purchasing x_{it} units of land at price p_{it} implies the following choice of investment in capital $W(k_{it}) - x_{it}p_{it}$, where $x_{it}p_{it}$ is the spending on the land market. Since the rate of return on capital is $f'(k_{it+1})$, it follows that the old agent's consumption bundle is $(c_{it+1}, A_i x_{it})$, where $c_{it+1} = [W(k_{it}) - x_{it}p_{it}]f'(k_{it+1}) + x_{it}p_{it+1}$. Thus the land market serves two purposes: First, land can be used as an investment vehicle, because it is durable and consumers can re-sell it in the next period in order to increase their consumption of manufacturing commodity. Second, land can be used as a consumption commodity, because land owners harvest the agricultural commodity from which they derive utility. Young agents choose the optimal levels of land/investment in capital by maximizing the following utility function $(c_{it+1}, x_{it}) \mapsto \ln c_{it+1} + u(A_i x_{it})$, where u is a twice continuously differentiable on $\mathbb{R}_{++}$; continuous, strictly increasing, and strictly concave on $\mathbb{R}_+$; and satisfies the following boundary condition, $\lim_{x \downarrow 0} u'(x) = \infty$.

Young agents take the pair of current - (k_{it}, p_{it}), and next period - (k_{it+1}, p_{it+1}), capital stock and land price as given and maximize the following lifetime utility

$$\ln \left\{ \left[W(k_{it}) - x_{it} p_{it} \right] f'(k_{it+1}) + x_{it} p_{it+1} \right\} + u(A_i x_{it}), \tag{1}$$

by choosing an optimal land holding, x_{it}. The first order condition implies that

(a)   If $p_{it} f'(k_{it+1}) > p_{it+1}$, then the optimal level of land holding, x_{it}, satisfies[h]

$$\frac{p_{it} f'(k_{it+1}) - p_{it+1}}{\left[W(k_{it}) - x_{it} p_{it} \right] f'(k_{it+1}) + x_{it} p_{it+1}} = A_i u'(A_i x_{it}). \tag{2}$$

(b)   If $p_{it} f'(k_{it+1}) \leq p_{it+1}$, then it is optimal for young agents to invest all their wage income into a land purchase and thus the optimal level of land holding is

$$x_{it} = \frac{W(k_{it})}{p_{it}}. \tag{3}$$

The autarky case

First let us consider the case of autarky, when all markets operate only domestically, foreign land ownership is not allowed, and price of land is determined domestically. A simple demographic structure of the consumption sector implies that young agents are net demanders, while old agents are net suppliers of land. Since all markets operate only domestically and the aggregate supply of land is constant and normalized to unity, it follows that $x_{it} = 1$, for all t. This, with the assumptions of no first period consumption and full depreciation of capital, implies that the capital stock in the next period is

$$k_{it+1} = W(k_{it}) - p_{it}. \tag{4}$$

Suppose $p_{it} f'(k_{it+1}) \leq p_{it+1}$, then $x_{it} = 1$ implies that $p_{it} = W(k_{it})$ and thus $k_{it+1} = 0$. This implies a contradiction with $p_{it} f'(k_{it+1}) \leq p_{it+1}$ because $\lim_{k_{it+1} \downarrow 0} f'(k_{it+1}) = \infty$. Suppose $p_{it} f'(k_{it+1}) > p_{it+1}$, then $x_{it} = 1$ with (2) implies that

$$p_{it+1} = (p_{it} - U(A_i) W(k_{it})) f'(k_{it+1}) \quad \text{where } U(A_i) := \frac{A_i u'(A_i)}{1 + A_i u'(A_i)}. \tag{5}$$

Equations (4) and (5) together imply that the evolution of (k_{it}, p_{it}) in a closed economy under perfect-foresight dynamics is described by the following time one map $M^i(.,.) : \mathbb{R}_+^2 \to \mathbb{R}_+^2$,

$$\begin{pmatrix} k_{it+1} \\ p_{it+1} \end{pmatrix} = M^i(k_{it}, p_{it}) = \begin{pmatrix} m_1(k_{it}, p_{it}) \\ m_2^i(k_{it}, p_{it}) \end{pmatrix}, \tag{6}$$

where

$$m_1(k, p) := W(k) - p \text{ and } m_2^i(k, p) := (p - U(A_i) W(k)) f' \left[m_1(k, p) \right]. \tag{7}$$

(6) and (7) imply that the steady state pair (k_i^*, p_i^*) satisfies the following system of equations

$$p_i = W(k_i) - k_i \text{ and } p_i = U(A_i) \frac{W(k_i) f'(k_i)}{f'(k_i) - 1}. \tag{8}$$

After eliminating the steady state price of land, p_i, from the above system and then re-arranging terms, we obtain that the steady state capital, k_i^*, should solve the following equation

$$\Psi(k_i) = U(A_i) \text{ where } \Psi(k) := \left(1 - \frac{k}{W(k)}\right)\left(1 - \frac{1}{f'(k)}\right). \tag{9}$$

In order to simplify the analysis and focus on the situation when there is a unique steady state in a closed economy, we make the following assumption.

Assumption 1. *Suppose that the production and utility functions satisfy*

(a) $k \mapsto \frac{k}{W(k)}$ *is strictly increasing on* $\mathbb{R}_{++}$ *and* $\lim_{k \downarrow 0} \frac{k}{W(k)} = 0$; .
(b) $x \mapsto x u'(x)$ *is strictly decreasing on* $\mathbb{R}_{++}$;

Assumption 1.(a), can be interpreted as a restriction on the elasticity of substitution between capital and labor inputs. In particular, Assumption 1.(a) is equivalent to $\sigma(k) > \alpha(k)$, where

$$\sigma(k) := \frac{W(k) f'(k)}{W'(k) f(k)} \text{ and } \alpha(k) := \frac{k f'(k)}{f(k)} \tag{10}$$

denote the elasticity of substitution between capital and labor inputs and capital share in production respectively. Justification for Assumption 1.(a) is based on empirical observations made on the one hand by Klump et al. (10), and Chirinko (4) who report that the elasticity of substitution between capital and labor inputs is between 0.5 and 1 and on the other hand by Gollin (7) who reports that capital share in production is between 0.32 to 0.40.

Assumption 1.(b), can be interpreted as a restriction on the income elasticity of demand for the agricultural commodity. In particular, Assumption 1.(b) implies that the income elasticity of demand for the agricultural commodity is less than one.[i] Justification for low income elasticity is based on the empirical observation of Engel's law made by Crafts (6), Matsuyama (12), and Piyabha et al. (14).

Let us define a constant

$$\widehat{k} := \min \{k \in \mathbb{R}_{++} | \Psi(k) = 0\}. \tag{11}$$

One can easily verify that if assumption 1.(a) is satisfied then (a) $k \mapsto \Psi(k)$ is strictly decreasing and positive on the interval $(0, \widehat{k})$, and (b) Ψ satisfies the following boundary conditions $\lim_{k \downarrow 0} \Psi(k) = 1$ and $\Psi(\widehat{k}) = 1$. At the same time, if assumption 1.(b) is satisfied then $A \mapsto U(A)$ is strictly decreasing and $U(A) \in (0, 1)$ for any $A \in (0, \infty)$.

Proposition 1. *If Assumption 1 is satisfied then (a) there exists a unique and interior steady state in a closed economy,* $(K^*(A_1), K^*(A_2))$; *and (b) for any initial capital stocks,* $(k_{10}, k_{20}) > 0$, *there exists a unique saddle path along which the economy converges monotonically to* $(K^*(A_1), K^*(A_2))$.

Proof of the above Proposition can be found in Appendix. The above proposition implies that, under autarky, there exists a unique and interior steady state to which the world economy converges monotonically along the unique saddle path. Convergence happens for any initial distribution of capital stock, (k_{10}, k_{20}). Under Assumption 1, the model predicts that the economy with more efficient agricultural productivity accumulated more capital[j] and achieves higher steady state welfare.[k] This result can be tempting to conclude that poor countries will benefit from foreign land ownership if such ownership comes with the transfer of more efficient agricultural technology.[l]

Two country model

The positive link between agricultural productivity and capital accumulation demonstrated above crucially depends on the closed economy assumption. To see this, consider a world economy in which the foreign land ownership is allowed. Suppose the ownership right comes with the transfer of agricultural productivity. Let efficiency in the agricultural sector, in both countries, be described by the same parameter A (we can assume without loss of generality that $A = A_2 > A_1$). On the one hand, capital accumulation in each country implies the following equations to hold[m]

$$k_{1t+1} = W(k_{1t}) - x_{1t}p_t \text{ and } k_{2t+1} = W(k_{2t}) - x_{2t}p_t. \tag{12}$$

On the other hand, it follows from agents' optimization and from the equalization of land prices that

$$p_{t+1} = \left(p_t - \frac{U(Ax_{it})}{x_{it}}W(k_{it})\right)f'(k_{it+1}) \tag{13}$$

to hold for $i = 1, 2$. (12) and (13) with land market clearing condition $x_{1t} + x_{2t} = 2$ implies that $x_{1t} = X(k_{1t}, k_{2t}, p_t)$ should solve the following equation

$$\frac{f'\left[W(k_{1t}) - x_{1t}p_t\right]}{f'\left[W(k_{2t}) - (2 - x_{1t})p_t\right]} = \frac{p_t - \frac{U(A(2-x_{1t}))}{2-x_{1t}}W(k_{2t})}{p_t - \frac{U(Ax_{1t})}{x_{1t}}W(k_{1t})}. \tag{14}$$

(12), (13), and (14) together implies that the evolution of the world economy under the perfect foresight dynamics is described by the following time one map

$$\begin{pmatrix} k_{1t+1} \\ k_{2t+1} \\ p_{t+1} \end{pmatrix} = N(k_{1t}, k_{2t}, p_t) = \begin{pmatrix} n_1(k_{1t}, k_{2t}, p_t) \\ n_2(k_{1t}, k_{2t}, p_t) \\ n_3(k_{1t}, k_{2t}, p_t) \end{pmatrix}, \tag{15}$$

where n_1, n_2, and n_3 are given by

$$n_1(k_1, k_2, p) := W(k_1) - X(k_1, k_2, p)p$$

$$n_2(k_1, k_2, p) := W(k_2) - (2 - X(k_1, k_2, p))p \tag{16}$$

$$n_3(k_1, k_2, p) := \left(p - \frac{U(AX(k_1, k_2, p))}{X(k_1, k_2, p)}W(k_1)\right)f'\left[n_1(k_1, k_2, p)\right].$$

Steady state analysis of the closed economy implies the existence of two symmetric steady states $(0, 0)$ and $(K^*(A), K^*(A))$ (in which both countries hold equal amounts of capital and incomes are equalized) and two asymmetric steady states $(0, K^*(2A))$ and $(K^*(2A), 0)$ (in which one country holds positive amount of capital, while the other deteriorates to zero level of capital). In order to investigate the existence of additional steady states, in which both countries hold positive but different levels of capital, we refine Assumption 1.

Assumption 2. *Suppose production and utility functions are:*

(a) $f(k) = k^\alpha$, *where the parameter* $\alpha \in (0,1)$ *describes the capital share in production;*
(b) $u(x) = \frac{x^{1-\gamma}-1}{1-\gamma}$, *where the parameter* $\gamma \in (1,\infty)$ *describes the income elasticity of demand on agricultural commodity;*

One can easily verify that Assumption 2 is fully compatible with Assumption 1. Thus, results obtained in Section The autarky case remains. It follows from (12), (13), and (14), that for a given steady state distribution of land holding $(x, 2-x)$, the steady state capital pair (k_1, k_2) solves the following system of equations

$$\Psi(k_1) = U(Ax) \text{ and } \Psi(k_2) = U(A(2-x)). \tag{17}$$

It follows from Assumption 2 and from expression (11) that $\widehat{k} = \min\{\alpha^{\frac{1}{1-\alpha}}, (1-\alpha)^{\frac{1}{1-\alpha}}\}$ and the distribution of the steady state capital stocks in the world economy is given by $k_1 = K^*(Ax)$ and $k_2 = K^*(A(2-x))$, where

$$K^*(y) := \left(\frac{1}{2} - \sqrt{\frac{1}{4} - \frac{\alpha(1-\alpha)}{1+y^{1-\gamma}}}\right)^{\frac{1}{1-\alpha}} \in (0, \widehat{k}). \tag{18}$$

Price equalization implies that the steady state land holding in the home country, x, should satisfy the following equation $\Delta(x, A) = 0$, where $\Delta : (0,2) \times \mathbb{R}_+ \to \mathbb{R}$ and $P^* : \mathbb{R}_{++} \to \mathbb{R}$ are defined as

$$\Delta(x, A) := P^*(Ax) - P^*(A(2-x)) \text{ and } P^*(y) = A\frac{U(y)}{y}\frac{W(K^*(y))f'(K^*(y))}{f'(K^*(y)) - 1}. \tag{19}$$

Proposition 2. *If* $\alpha\gamma > 1$ *then* $\frac{\partial\Delta(x,A)}{\partial x}\big|_{x=1} = 0$ *has a unique solution at* $A = A^c(\gamma)$, *where* A^c *is a strictly increasing function and satisfies the following boundary conditions*

$$A^c\left(\frac{1}{\alpha}\right) = 0 \text{ and } \lim_{\gamma \uparrow \infty} A^c(\gamma) = 1. \tag{20}$$

(a) *If either* $\alpha\gamma \leq 1$ *or* $\alpha\gamma > 1$ *and* $A \leq A^c(\gamma)$ *then there exists a unique and symmetric steady state in the world economy* $(K^*(A), K^*(A))$, *which is globally stable.*
(b) *If* $\alpha\gamma > 1$ *and* $A > A^c(\gamma)$ *then there exists three interior steady states with land holdings* $(x_L(A), x_H(A))$, $(1,1)$, *and* $(x_H(A), x_L(A))$, *where* $x_L(A) = 2 - x_H(A) \in (0,1)$ *solves* $\Delta(x, A) = 0$. *Symmetric steady state* $(K^*(A), K^*(A))$ *is unstable, while two asymmetric steady states*

$$(K^*[Ax_L(A)], K^*[Ax_H(A)]) \text{ and } (K^*[Ax_H(A)], K^*[Ax_L(A)]) \tag{21}$$

are asymptotically stable.

The above Proposition implies that the parameter space $(\gamma, A) \in (1, \infty) \times (0, \infty)$ can be divided into two regions (see Figure 1). When either $\alpha\gamma \leq 1$ or $\alpha\gamma > 1$ and $A \leq A^c(\gamma)$ then the world economy converges to a symmetric steady state and thus the relatively poor country with an inefficient agricultural sector will benefit from the land market liberalization in the long run, while the relatively rich country with more efficient agricultural sector will neither lose nor benefit from the land market liberalization.

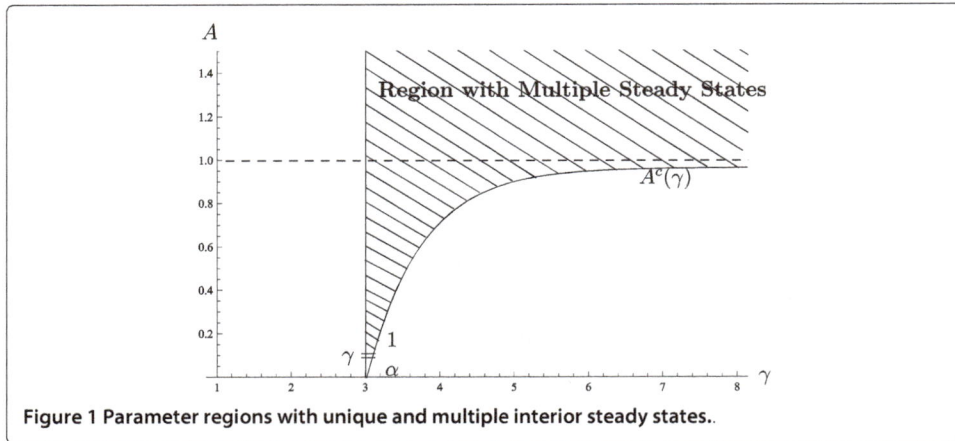

Figure 1 Parameter regions with unique and multiple interior steady states..

If $\alpha\gamma > 1$ and $A > A^c(\gamma)$ then the unique and symmetric steady state is unstable and two asymptotically stable steady states

$$(K^*[Ax_L(A)], K^*[Ax_H(A)]) \text{ and } (K^*[Ax_H(A)], K^*[Ax_L(A)]) \tag{22}$$

exist as only possible long-run outcomes in the world economy. The possibility for the existence of two asymmetric and interior steady states is shown in Figure 2. In such a situation, a country with a relatively inefficient agricultural sector can either benefit or lose from the land market liberalization. The long-run outcome depends on the relative levels of the agricultural productivity and the capital stock at the time of land market liberalization. In order to demonstrate the situation when a country finds itself in a disadvantage situation due to a land market liberalization, we consider the case when $\alpha\gamma > 1$ and $A = A_2 > A^c(\gamma)$. Suppose $A_1 \in (0, A)$, then the world economy under autarky will converge to the steady state $(K^*(A_1), K^*(A))$, where $0 < K^*(A_1) < K^*(A)$. Suppose the relatively poor country liberalizes the land market and the transfer of agricultural technology from rich to poor country occurs. Then the world economy will converge to the steady state in which the distribution of land holdings is $(x_1, x_2) = (x_L(A), x_H(A))$, and the distribution of capital is $(K^*[Ax_L(A)], K^*[Ax_H(A)])$. It is clear that the home country will find itself in a disadvantaged situation[n] due to the land market liberalization if $A_1 \in (Ax_L(A), A)$ and will benefit from the land market liberalization if $A_1 \in (0, Ax_L(A))$. This is visualized in Figure 3.

The main results obtained in the paper can be summarized by Figure 4. If either $\alpha\gamma \leq 1$ or $\alpha\gamma > 1$ and the productivity of the agricultural sector in the relatively rich country is sufficiently low, $A \leq A^c(\gamma)$, then the world economy will converge to a symmetric steady

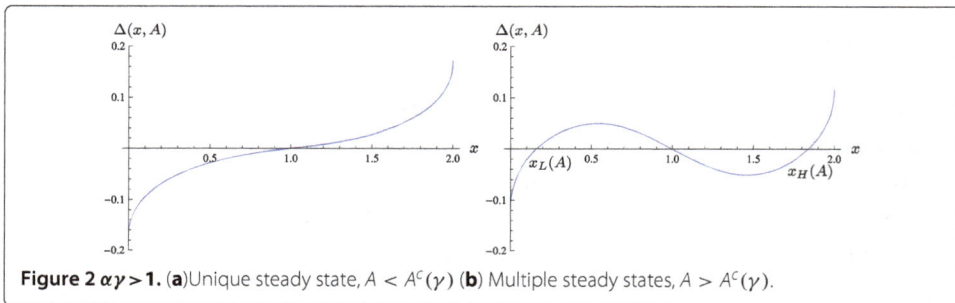

Figure 2 $\alpha\gamma > 1$. (a) Unique steady state, $A < A^c(\gamma)$ **(b)** Multiple steady states, $A > A^c(\gamma)$.

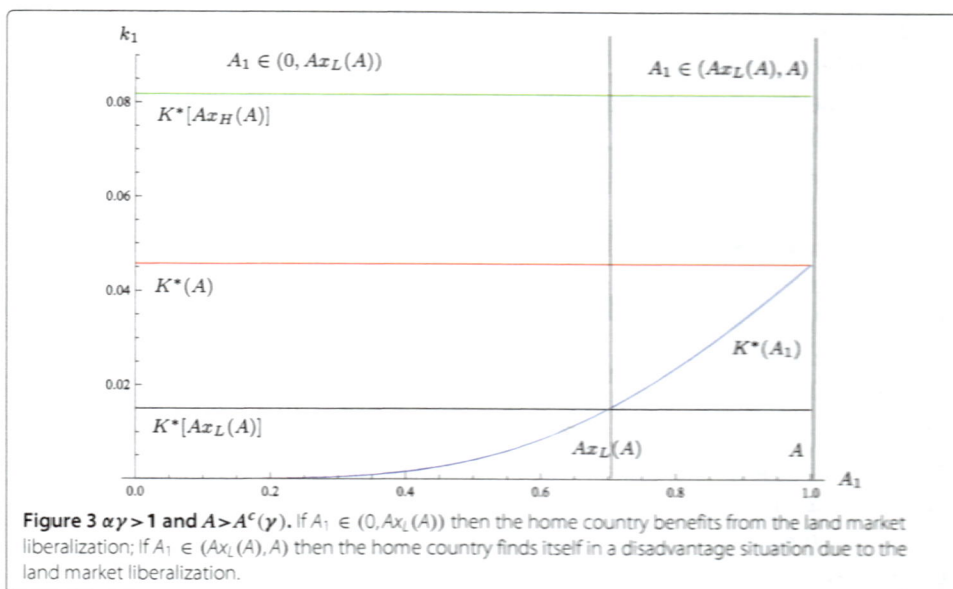

Figure 3 $\alpha\gamma > 1$ and $A > A^c(\gamma)$. If $A_1 \in (0, Ax_L(A))$ then the home country benefits from the land market liberalization; If $A_1 \in (Ax_L(A), A)$ then the home country finds itself in a disadvantage situation due to the land market liberalization.

state $(K^*(A), K^*(A))$ due to the land market liberalization. As a result, in the long run, only the relatively poor country benefits from the liberalization process, while the relatively rich country will sustain the same level of capital, income, and welfare in the long run (see region $\mathcal{NW}$, no-lose/win region, on Figure 4). If $\alpha\gamma > 1$ and the productivity of the agricultural sector is sufficiently high, $A > A^c(\gamma)$, then the world economy will be divided into relatively rich and relatively poor countries even when the productivity across countries is the same. Whether the land market liberalization helps or hurts the process of capital accumulation and the long-run welfare depends on the relative strength of the productivity improvement in the home country. In particular, if the productivity of the agricultural sector in the home country is sufficiently low, $A_1 < Ax_L(A)$, then both countries will benefit from the land market liberalization, because $0 < K^*(A_1) < K^*[Ax_L(A)]$

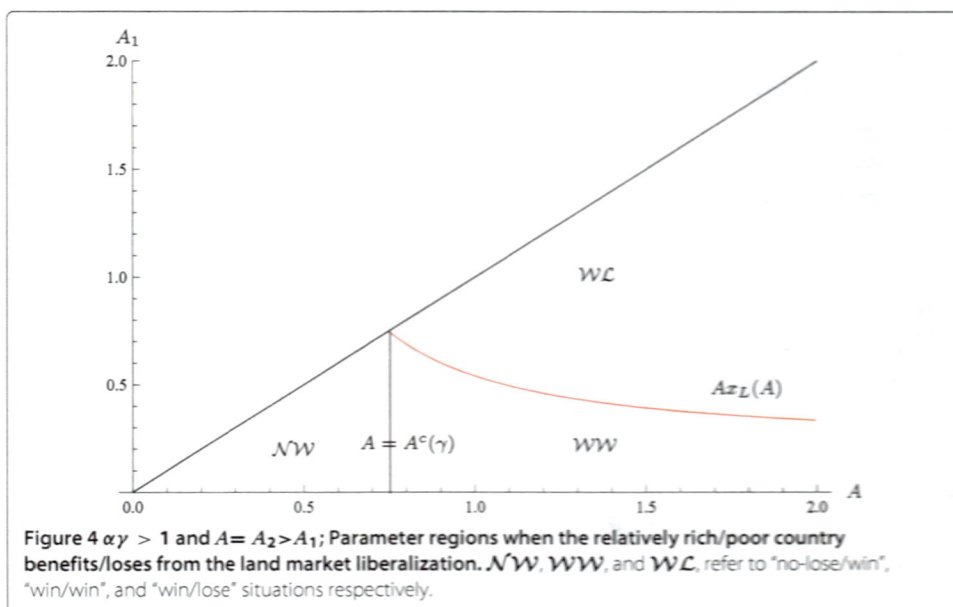

Figure 4 $\alpha\gamma > 1$ and $A = A_2 > A_1$; Parameter regions when the relatively rich/poor country benefits/loses from the land market liberalization. $\mathcal{NW}$, $\mathcal{WW}$, and $\mathcal{WL}$, refer to "no-lose/win", "win/win", and "win/lose" situations respectively.

and $0 < K^*(A) < K^*(Ax_H(A))$ (see region $\mathcal{WW}$, win-win region, on Figure 4). In contrast, if the productivity of the agricultural sector in the home country is sufficiently high, $A_1 > Ax_L(A)$, then the home country will suffer while the foreign country will benefit from the land market liberalization because $K^*(A_1) > K^*[Ax_L(A)]$ and $0 < K^*(A) < K^*[Ax_H(A)]$ (see region $\mathcal{WL}$, win-lose region, on Figure 4).

Results and discussion

The main goal of this paper was to analyze the effects of land market liberalization and the transfer of agricultural technology on the dynamics of capital accumulation and on long run welfare. The main results of the paper suggests that a policy maker should be careful to use only positive feedback mechanisms that the foreign land ownership can create for economic development, because in reality, several different links co-exist between the foreign land ownership right and the long run economic development. That is why an important lesson learned from this study is to be careful in any policy intervention; a clear goal to speed up the process of development can indeed push the country even deeper into an underdevelopment trap.

Conclusions

The paper considered a model of the world economy consisting of structurally identical economies and analyzed the stability of a unique symmetric steady state with and without foreign land ownership rights. This does not imply that initial differences in production technology, consumer preferences, or land endowment are unimportant for the world economy not to converge to the symmetric steady state. Instead, I demonstrated that even a small amount of inequality, expressed in terms of initial capital stock, can be magnified after foreign land ownership is allowed. I do not argue that foreign land ownership should be blamed for non-convergence and other sources of inequality such as geography, credit market imperfections, lack of transparency in corporate governance, and poorly designed public policies, are unimportant sources for non-convergence. Instead, I demonstrated that the foreign land ownership alone can create advantage/disadvantage for initially rich/poor countries and halt the process of convergence.

    The model was setup in such a way that the world economy in autarky converges to a unique steady state, which is globally stable under rational expectations. In such steady state, a country with a more productive agricultural sector accumulates more capital. Introduction of foreign land ownership rights puts a relatively rich/poor country in a advantageous/disadvantageous situation and creates an upward/downward spiral of high/low wage income, and high/low investment in domestic physical capital. The argument of the paper can be used to dispute the claim made in the World Bank annual report that a large-scale foreign land acquisition is a way to reduce poverty (see World Bank (17)). Instead, I argue that a large-scale foreign land acquisition in some cases can create a "win–win" for rich investors and a "lose–lose" outcome for poor, land selling countries.

Appendix

    Lemma 1. *Let the agent's utility be given by* $(c, x) \mapsto \log c + u(Ax)$. *If* $x \mapsto xu'(x)$ *is strictly decreasing then the income elasticity of agricultural commodity demand is less than unity.*

Proof. Let w and p denote current wage and land price, and let r_1 and p_1 denote next period rate of return on capital and next period land price. Then it follows (2) that demand of agricultural commodity satisfies the following equation

$$\frac{x(w,p)}{U(Ax(w,p))} = \frac{r_1}{pr_1 - p_1} w. \tag{23}$$

This implies that

$$\frac{d\ln x(w,p)}{d\ln w} = \frac{1}{1 - \frac{Ax(w,p)U'(Ax(w,p))}{U(Ax(w,p))}}, \tag{24}$$

where

$$\frac{yU'(y)}{U(y)} = \left(1 + \frac{yu''(y)}{u'(y)}\right) \frac{1}{1 + yu'(y)}. \tag{25}$$

$y \mapsto yu'(y)$ is strictly decreasing (see Assumption 1) if and only if $\frac{yu''(y)}{u'(y)} < -1$. This with (25) implies that $\frac{yU'(y)}{U(y)} < 0$, which with (24) implies that $\frac{d\ln x(w,p)}{d\ln w} \in (0,1)$. $\qquad\square$

Proof of Proposition 1

(a) An interior steady state of the closed economy i is a stationary pair (k_i^*, p_i^*) which solves (9).[o] The existence and uniqueness of the interior steady state $K^*(A_i)$ follows from (9) and from the following properties of Ψ: Ψ is strictly decreasing, $\Psi(\widehat{k}) = 0$, and $\lim_{k \downarrow 0} \Psi(k) = 1$.

(b) In order to understand the dynamics of the economy, I evaluate the Jacobian matrix of time one map at the unique and interior steady state $k_i^* = K^*(A_i)$. (4) and (5) imply that

$$m_{11}(k_i^*, p_i^*) = W'(k_i^*) \quad \text{and} \quad m_{12}(k_i^*, p_i^*) = -1, \tag{26}$$

and

$$m_{21}^i(k_i^*, p_i^*) = p_i^* W'(k_i^*) \left(\frac{f''(k_i^*)}{f'(k_i^*)} - \frac{f'(k_i^*)-1}{W(k_i^*)}\right)$$

$$m_{22}^i(k_i^*, p_i^*) = -p_i^* \frac{f''(k_i^*)}{f'(k_i^*)} + f'(k_i^*). \tag{27}$$

(26) and (27) together imply that the trace and determinant of the Jacobian matrix at the interior steady state are:

$$T_i^* = \frac{W'(k_i^*)}{f'(k_i^*)} \frac{f(k_i^*)-k_i^*}{k_i^*} + f'(k_i^*) \quad \text{and} \quad D_i^* = W'(k_i^*)\frac{f(k_i^*)-k_i^*}{W(k_i^*)}. \tag{28}$$

Since $f'(k_i^*) > 1$ and $W' > 0$, it follows from (28) that $T_i^* > 0$ and $D_i^* > 0$. In addition,

$$1 - T_i^* + D_i^* = 1 - f'(k_i^*) + f''(k_i^*)W(k_i^*) + f'(k_i^*)W'(k_i^*) = W(k_i^*)f'(k_i^*)\Psi'(k_i^*) < 0, \tag{29}$$

because $\Psi'(k_i^*) < 0$. (29) with $T_i^* > 0$ and $D_i^* > 0$ implies that a unique and interior steady state is a saddle. Local properties of the interior steady state implies the existence and uniqueness of a locally stable manifold, which is one-dimensional, continuously differentiable manifold, tangential to the linear space spanned by the eigenvector of the Jacobian matrix with eigenvalue less than one. Global properties of the dynamical system given in equation (4) can be established via the application of the Stable Manifold Theorem (see Hartman (9) and Guckenheimer and Holmes (8) for more details). Global invertibility of the time one map with uniqueness of an interior steady state implies the existence and global uniqueness of a perfect-foresight equilibrium which can be obtained through the backward iteration of the locally stable manifold. Existence of the globally stable manifold implies that for any k_{i0} there exists a unique p_{i0} such that the pair (k_{i0}, p_{i0}) is on the globally stable manifold. The economy converges to steady state equilibrium monotonically along the stable manifold. For any values other than p_{i0}, the economy steps on diverging path which is inconsistent with the perfect foresight equilibrium. The same argument is used in (Böhm V, Kikuchi T, Vachadze G: Natural resources and patterns of overtaking, Unpublished). □

Lemma 2. *Let us define a function* $Z : [0, \widehat{z}] \to \mathbb{R}$, *where* $\widehat{z} := \min\{\alpha, 1 - \alpha\}$, *as follows*

$$Z(z) := \frac{\gamma - 1}{\alpha(1-\alpha)^2} \frac{(\alpha(1-\alpha) - z)(\alpha - z)(1-z)}{1 - 2z}. \tag{30}$$

(a) *If* $\gamma \in \left(1, \frac{1}{\alpha}\right)$ *then* $Z(z) < 1$ *for* $z \in [0, \widehat{z}]$;
(b) *If* $\gamma = \frac{1}{\alpha}$ *then* $Z(0) = 1$ *and* $Z(z) < 1$ *for* $z \in (0, \widehat{z}]$;
(c) *If* $\gamma \in \left(\frac{1}{\alpha}, \infty\right)$ *then* $Z(z) = 1$ *has a unique solution on the interval* $(0, \widehat{z})$;

Proof. It is easy to observe that $Z(\alpha(1-\alpha)) = 0$ and

$$Z(0) = \frac{\alpha(\gamma - 1)}{1 - \alpha} \in \begin{cases} (0, 1) & \text{if } \alpha\gamma < 1 \\ 1 & \text{if } \alpha\gamma = 1 \\ (1, \infty) & \text{if } \alpha\gamma > 1. \end{cases} \tag{31}$$

If $\alpha \in (0, 0.50]$, then $\widehat{z} = \alpha$, $Z(\alpha) = 0$, and Z is strictly decreasing and positive on $(0, \alpha(1-\alpha))$. If $\alpha \in (0.50, 1)$, then $\widehat{z} = 1 - \alpha$ and $Z(1-\alpha) = 1 - \gamma < 0$, and Z is strictly decreasing and positive on $(0, \alpha(1-\alpha))$. This implies the claim of the lemma. □

□

Lemma 3. (a) *If* $\alpha\gamma < 1$ *then* P^* *is strictly decreasing and satisfies the boundary conditions* $\lim_{y \downarrow 0} P^*(y) = \infty$ *and* $\lim_{y \uparrow \infty} P^*(y) = 0$;
(b) *If* $\alpha\gamma = 1$ *then* P^* *is strictly decreasing and satisfies the boundary conditions* $\lim_{y \downarrow 0} P^*(y) = \alpha^{\frac{2\alpha-1}{1-\alpha}}(1-\alpha)^{\frac{\alpha}{1-\alpha}}$ *and* $\lim_{y \uparrow \infty} P^*(y) = 0$;
(c) *If* $\alpha\gamma > 1$ *then there exists a unique* y^c *which solves the equation*

$$\frac{\gamma - 1}{\alpha(1-\alpha)^2} \frac{(\alpha(1-\alpha) - \left[K^*(y)\right]^{1-\alpha})(\alpha - \left[K^*(y)\right]^{1-\alpha})(1 - \left[K^*(y)\right]^{1-\alpha})}{1 - 2\left[K^*(y)\right]^{1-\alpha}} = 1. \tag{32}$$

P^* *is strictly increasing on the interval* $(0, y^c)$, *strictly decreasing on the interval* (y^c, ∞) *and satisfies the boundary conditions* $\lim_{y \downarrow 0} P^*(y) = \lim_{y \uparrow \infty} P^*(y) = 0$;

Proof. It follows from (18) and (19) that

$$\frac{dK^*(y)}{dy} \frac{y}{K^*(y)} = \frac{\gamma - 1}{\alpha(1-\alpha)^2} \frac{(1 - \alpha - \left[K^*(y)\right]^{1-\alpha}(y))(\alpha - \left[K^*(y)\right]^{1-\alpha})(1 - \left[K^*(y)\right]^{1-\alpha})}{1 - 2\left[K^*(y)\right]^{1-\alpha}}$$

$$\tag{33}$$

and

$$\frac{dP^*(y)}{dy}\frac{y}{P^*(y)} = \frac{\gamma-1}{\alpha(1-\alpha)^2}\frac{(\alpha(1-\alpha)-[K^*(y)]^{1-\alpha})(\alpha-[K^*(y)]^{1-\alpha})(1-[K^*(y)]^{1-\alpha})}{1-2[K^*(y)]^{1-\alpha}}$$
$$-1.$$

(34)

(34) with Lemma 2 implies the monotonicity properties of P^*. We can observe that K^* is strictly increasing with

$$\lim_{y\downarrow 0} K^*(y) = 0 \quad\text{and}\quad \lim_{y\uparrow\infty} K^*(y) = \widehat{k}.$$

(35)

(35) implies that, for any $\gamma \in (1,\infty)$, $\lim_{y\uparrow\infty} P^*(y) = 0$ and

$$\lim_{y\downarrow 0} P^*(y) = \frac{\gamma-1}{1-\alpha}(\alpha(1-\alpha))^{\frac{\alpha}{1-\alpha}}y^{\frac{\alpha\gamma-1}{1-\alpha}} = \begin{cases} \infty & \text{if } \alpha\gamma < 1 \\ \alpha^{\frac{2\alpha-1}{1-\alpha}}(1-\alpha)^{\frac{\alpha}{1-\alpha}}\alpha & \text{if } \alpha\gamma = 1 \\ 0 & \text{if } \alpha\gamma > 1. \end{cases}$$

(36)

□

Proof of Proposition 2 Monotonicity of A^c follows from Lemmas 3 and 2 and from expression (32). The same Lemmas also imply that (a) if $\alpha\gamma = 1$ then $K^*(A^c(\gamma)) = 0$ and thus $A^c(\gamma) = 0$; and (b) if $\gamma \uparrow \infty$ then $[K^*(A^c(\gamma))] \uparrow \alpha(1-\alpha)$ and thus $A^c(\gamma) \uparrow 1$ because $U(A)$ and $K^*(A)$ converges pointwise to $U_\infty(A)$ and $K^*_\infty(A)$ respectively as $\gamma \uparrow \infty$, where

$$U_\infty(A) = \begin{cases} 1 & \text{if } A < 1 \\ 0.50 & \text{if } A = 1 \\ 0 & \text{if } A > 1, \end{cases} \quad\text{and}\quad K^*_\infty(A) = \begin{cases} 0 & \text{if } A < 1 \\ K^*(1) & \text{if } A = 1 \\ \widehat{k} & \text{if } A > 1. \end{cases}$$

(37)

Steady States: It follows from Lemma 3 that if $\alpha\gamma < 1$ then $\Delta(.,A)$ is continuous, strictly decreasing, and satisfies the boundary conditions

$$\lim_{x\downarrow 0} \Delta(x,A) = \infty \quad\text{and}\quad \lim_{x\uparrow 2} \Delta(x,A) = -\infty.$$

(38)

This implies that $\Delta(x,A) = 0$ admits a unique solution at $x = 1$. If $\alpha\gamma > 1$ then

$$\lim_{x\downarrow 0} \Delta(x,A) = -P^*(2A) < 0 \quad\text{and}\quad \lim_{x\uparrow 2} \Delta(x,A) = P^*(2A) > 0.$$

(39)

This implies that $\Delta(x,A) = 0$ can admits at most three solutions, $x_L(A) \in (0,1)$, $x_M(A) = 1$, and $x_H(A) = 2 - x_L(A) \in (1,2)$. If $\alpha\gamma > 1$ then it follows from Lemma 2 that

$$\frac{(\alpha(1-\alpha)-[K^*(A)]^{1-\alpha})(\alpha-[K^*(A)]^{1-\alpha})(1-[K^*(A)]^{1-\alpha})}{1-2[K^*(A)]^{1-\alpha}} = \frac{\alpha(1-\alpha)^2}{\gamma-1},$$

(40)

admits a unique solution. Let $A = A^c(\gamma)$ denotes the solution. Lemmas 3 and 2 imply that if $A < A^c(\gamma)$ then $\Delta(.,A)$ is increasing at $x = 1$ and thus there exists a unique solution at $x = 1$. In contrast, if $A > A^c(\gamma)$ then $\Delta(.,A)$ is decreasing at $x = 1$ and thus $\Delta(x,A) = 0$ admits three solutions. □

Stability: $\alpha\gamma > 1$ and $A > A^c(\gamma)$ is a necessary and sufficient condition for instability of symmetric steady state.

Sufficiency: In order to show that $\alpha\gamma > 1$ and $A > A^c(\gamma)$ is a sufficient condition for instability of symmetric steady state it is enough to show that the dynamical system (13) is locally unstable at $k_1 = k_2 = K^*(A)$. This is equivalent to show that at least two roots of the characteristic equation

$$Det \begin{pmatrix} n_{11} - \lambda & n_{12} & n_{13} \\ n_{12} & n_{11} - \lambda & n_{13} \\ n_{31} & n_{31} & n_{33} - \lambda \end{pmatrix} = 0, \tag{41}$$

are more than unity in modulus. Characteristic roots of the Jacobian matrix solve the following equation

$$(n_{11} - \lambda)^2 (n_{33} - \lambda) + 2n_{12}n_{13}n_{31} - 2(n_{11} - \lambda) n_{13}n_{31} - (n_{33} - \lambda) n_{12}^2 = 0. \tag{42}$$

Grouping first and fourth and second and third terms together, we obtain that the characteristic roots satisfy the following equation

$$(n_{11} - n_{12} - \lambda)(n_{11} + n_{12} - \lambda)(n_{33} - \lambda) - 2(n_{11} - n_{12} - \lambda) n_{13}n_{31} = 0. \tag{43}$$

It follows from the above expression that

$$\lambda_1 = n_{11} - n_{12}, \tag{44}$$

while λ_2 and λ_3 satisfy the following quadratic equation

$$\lambda^2 - \lambda (n_{11} + n_{12} + n_{33}) + (n_{11} + n_{12}) n_{33} - 2n_{13}n_{31} = 0. \tag{45}$$

At symmetric steady state $k_1 = k_2 = k^*$, $x_1 = x_2 = 1$, and thus it follows from (14) that

$$X_1(k^*, k^*, p^*) = -X_2(k^*, k^*, p^*) = \frac{W'(k^*)}{2} \frac{\frac{f'(k^*)}{W(k^*)} - \frac{f''(k^*)}{f'(k^*)-1}}{f'(k^*)\left(1 - \frac{AU'(A)}{U(A)}\right) - \frac{p^* f''(k^*)}{f'(k^*)-1}}, \tag{46}$$

and $X_3(k^*, k^*, p^*) = 0$. This with (17) and (44) implies that

$$\lambda_1 = W'(k^*) \frac{\frac{k^*}{W(k^*)} - \frac{AU'(A)}{U(A)}}{1 - \frac{p^*}{f'(k^*)} \frac{f''(k^*)}{f'(k^*)-1} - \frac{AU'(A)}{U(A)}}, \tag{47}$$

Since $-\frac{AU'(A)}{U(A)} = (\gamma - 1)(1 - U(A))$ and $\Psi(k^*) = U(A)$, it follows from (47) that

$$\lambda_1 > 1 \Leftrightarrow \frac{\gamma-1}{\alpha(1-\alpha)^2} \frac{(\alpha(1-\alpha)-[k^*]^{1-\alpha})(\alpha-[k^*]^{1-\alpha})(1-[k^*]^{1-\alpha})}{1-2[k^*]^{1-\alpha}} > 1. \tag{48}$$

This with (34) and with Lemmas 3 and 2 implies $\lambda_1 > 1$ if and only if $\alpha\gamma > 1$ and $A > A^c(\gamma)$. In addition,

$$\lambda_2 + \lambda_3 = n_{11} + n_{12} + n_{33} = W'(k^*) + f'(k^*) - p^* f''(k^*) > 0, \tag{49}$$

$$\lambda_2 \lambda_3 = (n_{11} + n_{12}) n_{33} - 2n_{13}n_{31} = W'(k)f'(k) > 0, \tag{50}$$

$$(1 - \lambda_2)(1 - \lambda_3) = 1 - f'(k) + \left[W(k)f'(k)\right]' < 0. \tag{51}$$

It follows from (49), (50), and (51), that both roots λ_2 and λ_3 are positive, real, and satisfy $0 < \lambda_2 < 1 < \lambda_3$. I.e., $\lambda_1 \geq 1$ is a sufficient condition for local instability of the symmetric steady state (k^*, k^*, p^*). Since there can exist at most two interior asymmetric steady states in the world economy, the local instability of the symmetric steady state should imply the long run divergence.

Necessity: In order to show that $\alpha\gamma > 1$ and $A > A^c(\gamma)$ is a necessary condition for a long run divergence it is enough to show that whenever divergence happens then $\alpha\gamma > 1$

and $A > A^c(\gamma)$ is satisfied. Here the fact that there are at most three interior steady states plays a crucial role. In particular, when there are more than three interior steady states then the symmetric steady state can be locally stable but the divergence in the world economy can still occur. However, when the number of interior steady states is at most three then the local and global instability becomes equivalent.

Endnotes

[a]Among these countries are: Argentina, Brazil, Bulgaria, Chile, Costa Rica, Georgia, Guatemala, Guyana, Latvia, Madagascar, Poland, Romania, Thailand, and others.

[b]Recent scientific debate on land speculation (or land grabbing) can be found in a special issue of *Journal of Peasant Studies* published in 2011 (Vol. 38, issue 2).

[c]According to FNP Institute, an agribusiness consulting firm in Sao Paulo, in 2007 alone, farmland prices jumped by 16% in Brazil, by 31% in Poland, and by 15% in the Midwestern United States.

[d]Two key assumptions behind this result is the diminishing returns technology and low income elasticity of agricultural demand.

[e]For details about the standard neoclassical overlapping generations model see Azariadis (2).

[f]Capital can be interpreted broadly to include human capital, physical capital, or any other capital good used in production.

[g]Matsuyama (12) analyses the effect of agricultural productivity on economic growth and shows that high agricultural productivity is a precondition for high economic growth in case of autarky, while high agricultural productivity can become a barrier for economic growth in case of free trade. This is so because, in case of autarky, rising agricultural productivity makes it possible to free labor force from the agricultural sector. However, in the case of free trade, high agricultural productivity causes the specialization in the agricultural sector and leads to de-industrialization. In order to separate this feedback mechanism from the one considered in this paper, we assume that the land is the only factor for agricultural commodity production.

[h] Properties of f and u imply the existence and uniqueness of x_{it} which solves (2).

[i]See Lemma 1 in the Appendix.

[j]Since $k \mapsto \Psi(k)$ and $A \mapsto U(A)$ are both decreasing functions it follows that $A \mapsto K^*(A)$, where $k = K^*(A)$ solves of $\Psi(k) = U(A)$, is a strictly increasing function.

[k]Ranking of steady state welfare across countries coincides with the ranking of steady state capital because $A \mapsto K^*(A)$ and $A \mapsto \ln(f(K^*(A)) - K^*(A)) + u(A)$ are both strictly increasing functions.

[l]This result is consistent with Matsuyama (12). Engel's law plays a crucial role for this result. If income elasticity is unity, then the steady state capital is independent from productivity in the agricultural sector. If income elasticity is more than one, and so agricultural commodity is a luxury good, then a rise in agricultural productivity decreases the steady state capital.

[m] It is assumed that the young agents cannot invest in capital abroad. This assumption can be easily justified if one interprets capital as human or public capital.

[n]As above, ranking of steady state welfare coincides with the ranking of steady state capital because $A \mapsto \ln(f(K^*(A)) - K^*(A)) + u(A)$ is a strictly increasing function.

[o]It is clear that $(0, 0)$ is a corner steady state of the economy.

Competing interests
The author declares that he has no competing interest.

Acknowledgements
This paper was partly written at the European Commission, Directorate-General for Economic and Financial Affairs, where I was a Visiting Fellow in April 2010. I would like to thank Peter Grasmann, Heikki Oksanen, and two anonymous referees for their helpful comments and suggestions.

References
1.   Allen TJ: **Global Land Grab.** *In These Times* 2011:1–2. http://inthesetimes.com/article/11784/global_land_grab/.
2.   Azariadis C: *Intertemporal macroeconomics*. London: Blackwell; 1993.
3.   Borras M, Hall R, Scoones I, White B, Wolford W: **Towards a better understanding of global land grabbing: an editorial introduction.** *J Peasant Stud* 2011, **38:**209–216.
4.   Chirinko RS: σ**: The long and short of it.** *J Macroeconomics* 2008, **30:**671–686.
5.   Cotula L, Vermeulen S, Leonard R, Keeley J: *Land grab or development opportunity? Agricultural investments and international land Delas in Africa, International institute for environment and development (IIED)/food and agriculture organization of the United Nations (FAO)/international fund for agricultural development (IFAD)*. London/Rome; 2009. http://www.fao.org/docrep/011/ak241e/ak241e00.htm.
6.   Crafts N: **Income elasticities of demand and the release of labour by agriculture during the British industrial revolution.** *J Eur Econ Hist* 1980, **9:**153–168.
7.   Gollin D: **Getting income shares right.** *J Pol Econ* 2002, **110:**458–74.
8.   Guckenheimer J, Holmes P: *Nonlinear oscillations, dynamical systems, and bifurcations of vector fields*. New York: Springer; 1983.
9.   Hartman P: *Ordinary differential equations*. New York: Wiley; 1964.
10. Klump R, McAdam P, Willman A: **Factor substitution and factor augmenting technical progress in the United States: a normalized supply-side system approach.** *Rev Econ Stat* 2007, **89:**183–192.
11. Li TM: **Centering labor in the land grab debate.** *J Peasant Stud* 2011, **38:**281–298.
12. Matsuyama K: **Agricultural productivity, comparative advantage, and economic growth.** *J Econ Theory* 1992, **58:**317–334.
13. Nierenberg D, Pollack B: **Innovations in access to land: land grab or agricultural investment?** *Huffington Post* 2010:1. http://www.huffingtonpost.com/danielle-nierenberg/innovations-in-access-to_b_671773.html.
14. Piyabha K, Rebelo S, Xie D: **Beyond balanced growth.** *Rev Econ Stud* 2001, **68:**869–882.
15. Vidal J: **Fears for the world's Poor countries as the rich grab land to grow food.** *Guardian* 2009:1. http://www.guardian.co.uk/environment/2009/jul/03/land-grabbing-food-environment.
16. Von Braun J, Meinzen-Dick R: *Land grabbing by foreign investors in Developing countries: Risks and opportunities. Policy brief: International food policy research institute*; 2009. http://www.ifpri.org/publication/land-grabbing-foreign-investors-developing-countries.
17. World Bank: *World Development Report 2008: Agriculture for development*. Washington, DC; 2008. pp 1–355. http://siteresources.worldbank.org/INTWDR2008/Resources/WDR_00_book.pdf.

2

Valuation of traits of indigenous sheep using hedonic pricing in Central Ethiopia

Zelalem G Terfa[1*], Aynalem Haile[2], Derek Baker[3] and Girma T Kassie[4]

* Correspondence: zedgutu@gmail.com
[1]The University of Liverpool, Liverpool, UK
Full list of author information is available at the end of the article

Abstract

This study estimates the implicit prices of indigenous sheep traits based on revealed preferences. A hedonic pricing model is fitted to examine the determinants of observed sheep prices. Transaction data were generated from rural markets of Horro-Guduru Wollega Zone of Ethiopia. Both OLS and heteroscedasticity consistent estimations were made. The empirical results consistently indicate that phenotypic traits of traded indigenous sheep (age, color, body size, and tail condition) are major determinants of price implying the importance of trait preferences in determining the price of sheep in local markets. Season and market locations are also very important price determinants suggesting the need to target season and market place in sheep improvement programmes. Therefore, the development of a comprehensive breeding program that has marketing element is crucial to make sheep improvement sustainable and sheep keepers benefit from the intervention.

Keywords: Hedonic pricing; Heteroscedasticity consistent; Phenotypic; Indigenous; Trait preference

Background

Small ruminants are a key component of the rural livelihood systems in rural Ethiopia. It is estimated that in 2010 Ethiopia owned about 48 million small ruminants (FAOSTAT, 2010) and this is one of the largest populations in sub Saharan Africa (SSA). Small ruminants contribute substantially to income, food (meat and milk), and non-food products like manure, skins and wool. They also serve as part of the crop failure risk coping portfolio of enterprises, for asset wealth security as form of money saving and investment as well as many other cultural functions (Tibbo, 2006). At farm household level, sheep contribute up to 63% of the net cash income derived from livestock production in the crop-livestock production systems in Ethiopia. In the dry low-lands of the country, sheep play a key role in sustaining the livestock-based pastoral and agro-pastoral livelihoods (Negassa and Jabbar, 2008). Despite the pronounced importance of small ruminants in general and sheep in particular, the productivity of the animals per head is considerably low. FAO (2009) estimated the average annual off-take rate and carcass weight per slaughtered animal for the period 2000 to 2007 to be 32.5% and 10.1 kg, respectively, the lowest even among the sub Saharan African countries. In fact, in the highlands of the country, sheep off-take was found to be even lower at 7% (Negassa and Jabbar, 2008).

In the study area (Horro-Guduru Wollega Zone of Ethiopia) sheep are equally important in the rural economy. The sheep production systems in the area, however, are traditional and semi-subsistence oriented. So far, only very limited efforts have been exerted to introduce and promote market-oriented sheep production and hence the current income generating capacity of the sector is not at all justifiable. Re-orientation of the production system, which involves designing an effective and informed breeding programme, is a necessity to bring about improvements in productivity and in the production system of the sector. This re-orientation entails proper valuation of both traded and non-traded products and services generated from the system. Information on the economic value of populations, traits and processes would ease the management of animal genetic resources that requires many decisions (Scarpa et al. 2003). Proper identification and valuation of the different characteristics would make resource allocation decisions among the different livestock improvement interventions for commercialization of the system quite fast and smooth (Drucker et al. 2001; Kassie, 2007). This will also enable identification of sheep market opportunities by identifying preferred traits of sheep. This is crucial as consumers' demand and preference is continuously changing over time.

Researchers have applied different economic valuation methods to understand the preference for and the value of animal traits in different contexts. Revealed preference and stated preference based models are the two most commonly used approaches. Revealed preferences based valuation methods record and analyze actual payments on observable transactions for the commodities/services of interest while stated preference based valuation methods make use of data on hypothetical choices and implicit payments (Hensher et al. 2005). Richards and Jeffrey (1996) employed a hedonic pricing model to establish indices of genetic worth of a dairy bull in Alberta, Canada. Their study indicated that the most important factors used by dairy farmers in valuing dairy bulls are milk volume, protein and fat content, general conformation, body capacity, and popularity of the bull. Barret et al. (2003) used a structural-heteroskedasticity-in-mean estimation method to identify the determinants of livestock producer prices in the dry lands of northern Kenya. Their result shows the importance of animal characteristics, periodic events that shift local demand or supply, and rainfall in determining prices producers receive. Williams et al. (2006), similarly used a hedonic model using weekly sales transactions to analyze cattle prices in West Africa and reported that location, season, and cattle attributes influence sheep price.

In their study that aimed at investigating determinants of inter-annual price variation of small ruminants' price in the eastern highlands of Ethiopia, Gezahegn et al. (2006) employed hedonic price modeling and reported significant differences in prices between seasons and markets, controlling for attributes of animals. Kassie et al. (2011a) similarly applied heteroscedasticity consistent hedonic price modeling to examine factors that influence cattle prices in the rural markets of central Ethiopia. The results of this study showed that season, market location, age, sex and body size are very important determinants of cattle price. Chang et al. (2010) employed hedonic price modeling to study price differentials of retailed eggs and reported significant premiums attributed to production method, variation in geographic locations and egg color. Similarly, Satimanon and Weatherspoon (2010) employed the same approach to determine price premiums of traits of fresh eggs using sustainable attribute data from retail markets in the United States. Their study indicates that welfare-managed eggs have a significant price premium while the sustainable packaging attributes are insignificant.

Other studies used a combination of revealed and stated preference data (e.g., Scarpa et al. 2003; Kassie et al. 2011b). Stated preference based valuation of animal genetic resources has also been widely used (e.g., Omondi et al. 2008; Kassie et al. 2009, and Faustin et al. 2010). In recent years there is a growing interest in using stated preference approaches which specifically employ choice experiments as real choice data in actual market are hardly available. Whenever available, however, revealed preference data have obvious advantages over stated preference data. Real world representation, embodiment of real constraints, reliability and validity are advantages of revealed preference data (Haab and McConnel, 2002; Hensher et al. 2005).

This brief review has shown that there is an enormous body of knowledge on the relevance and application of hedonic price models. Although the focus of most of the studies is market oriented production systems, the importance of the attributes of livestock in determining prices observed in the market is a key lesson to learn. Interestingly though, there are hardly any publications done on sheep price modeling in subsistence and/or semi-subsistence crop livestock mixed farming systems. This research employs the well-established hedonic price modeling in a context where markets are yet to develop and sheep have a more complex role than serving simply as sources of meat or in some cases wool.

Methods

The study area and the rural markets

The study was conducted in the Horro-Guduru Wollega zone of Ethiopia. The administrative capital of the zone is called Shambu and is located at about 310 km west of Addis Ababa. The 2007 population and housing census of the Central Statistical Agency showed that in 2007 the total population of the zone was about 580,000 out of which 50.1% were male and 49.9% were female (CSA, 2007). About 89% of the population in the zone lived in rural areas. The total area of the zone is about 710,000 hectares. According to the agency's national agricultural survey, the livestock population of the zone encompassed 127,000 heads of cattle, 25,000 sheep, and 12,000 goats (CSA, 2009).

The study covered four sheep markets, namely, the markets of *Shambu, Gaba Sanbata, Harato,* and *Fincha*. All the markets, except *Shambu*, are weekly markets that set on once in a week on a designated day. *Shambu* operates throughout the week, except on Sundays. The market infrastructure in the zone is very poor and there is no fence or shed, information provision, and feed provision in these four markets. *Fincha* is the only fenced sheep market where livestock are traded in a relatively organized manner. Livestock are trekked to and from all the markets. The study markets are among the remote rural markets dominated by (crop and livestock) farmers, farmer-traders, peri-urban butchers and small restaurant owners.

In these markets, grading and standardization do not exist and transactions take place after a long one-to-one bargaining between sellers and buyers on a per-head basis. The price paid by the buyer and received by the seller, therefore, depends, among others, on how well he or she can bargain. Under such circumstances, prices paid will reflect buyers' preference for various sheep traits, the type of buyer and seller and characteristics of the market place. The identification and analysis of the preferred traits and household characteristics that influence the prices actually paid in the market accordingly forms the basis for effective market development interventions. This study generated primary data and analyzed the factors that determine sheep prices in rural parts of the zone for this particular purpose.

The data

Data on 195 traded sheep and on sheep marketers' attributes were collected in the four rural sheep markets mentioned above. The main traits of traded sheep we focused on were coat color, body size, tail condition, age, and sex. Markets in developing countries in general and in such rural setups in particular are hardly competitive due to the sources of inefficiency mentioned above and all other generic sources of market imperfections. This entails the inclusion of factors apart from the attributes of the goods and services – in this case the sheep – in the model specification (Abdulai 2000; Kassie et al. 2011a). Therefore, we have generated and analyzed data on other factors that are expected to affect sheep price. These factors include the attributes of buyers and sellers, such as occupation and education level to serve as proxies for bargaining power. Seasonality of demand and supply was also captured. Description of variables used in this study is presented in Table 1 below.

Three survey rounds of individual sheep level transactions were conducted over an interval of one month. The first round was conducted during the beginning of January 2009; i.e., the Ethiopian Christmas season. This round was targeted to capture the price change that occurs during holidays. The second round was done in February 2009. This is a period with no important festival or planned social occasion and it overlaps with the time when farmers have completed crop harvesting. By this time farmers are expected to be less forced to sell their livestock for generating liquid capital (Kassie et al. 2011a). The third round was undertaken in March 2009. This period corresponds to the Ethiopian lent. From each market, 15 buyers were considered in each round except in *Harato* where 20 buyers were interviewed, taking into account the relative size of the market. That means 65 buyers were interviewed in each round.

The descriptive statistics of the variables used in the econometric model show considerably variation across respondents (Table 2). The average age of marketed sheep in the surveyed rural markets was one year (st.dev. = 11.4 months). In the observed sheep transactions in the four markets, 59% of traded sheep were male implying that female sheep are less frequently marketed as they are usually kept for reproduction (herd replacement) and less for generating cash. The average sheep price during this study was more than Ethiopian Birr 238.00. Sheep in those markets were observed to have different patterns of fur color although red was the dominant color (44%) followed by creamy-white (29%) over the survey period and observed transactions. More than 10% of the marketed sheep were black while the remaining were brown, white and mixed colored sheep. Data on body condition of the marketed sheep, which indicates relative fatness and appearance, were also observed and it was found that 48% of the sheep in the markets were in good condition and another 48% had a medium body condition. The remaining 3.6% were sheep with bad body condition. Related is the body size of the sheep marketed and 43% of them were medium sized, 33% small and the rest large size. The dominant tail condition of the traded sheep were long and thin tail type (48%) followed by long and fat tail type (24%). Most actors in the local sheep markets

Table 1 Summary of variables and coding method used in hedonic price model

Attribute	Code	Attribute	Code
Color		**Market place**	
White	1 = white	Gaba sanbata	1 = Gaba sanbata
	−1 = red		−1 = Shambu
	0 = otherwise		0 = otherwise
Black	1 = black	Fincha	1 = Fincha
	−1 = red		−1 = Shambu
	0 = otherwise		0 = otherwise
Brown	1 = brown	Harato	1 = Harato
	−1 = red		−1 = Shambu
	0 = otherwise		0 = otherwise
Creamy white	1 = creamy white	**Seller type**	
	−1 = red	Farmer	1 = farmer
	0 = otherwise		−1 = trader
White mixed	1 = white mixed		0 = otherwise
	−1 = red	Farmer trader	1 = farmer trader
	0 = otherwise		−1 = trader
Sex of sheep	0 - female		0 = otherwise
	1 - male	**Buyer type**	
Body size		Trader	1 = buyer is trader
Medium size	1 = medium		−1 = other buyers
	−1 = small		0 = otherwise
	0 = otherwise	Farmer	1 = buyer is farmer
Big	1 = big		−1 = other buyers
	−1 = small		0 = otherwise
	0 = otherwise	Farmer trader	1 - buyer is farmer trader
Tail type			−1 = other buyers
Medium and thin	1 = medium and thin		0 = otherwise
	−1 = long and fat	**Season**	
	0 = otherwise	Christmas (season 1)	1 = Christmas
Medium and fat	1 = medium and fat		−1 = season 2
	−1 = long and fat		0 = otherwise
	0 = otherwise	Fasting (season 3)	1 = fasting season
			−1 = season 2
			0 = otherwise

were farmers and sheep traders. Though data on the qualitative attributes were entirely based on buyers' perception of traded sheep, it is important to understand buyers' preference for these attributes. Generally, the typical sheep traded in these markets is 12 months old, red coated, of good or medium body condition, medium body size, and long thin tailed.

Analytical framework
Revealed preference is manifested through the actual prices paid for goods and services with expected utility. Hence, the prices sheep sellers receive are reflections of the utility

Table 2 Descriptive statistics for sheep in the local market and market participants

Description	Mean(SD)/percentage
Price per head of sheep (ETB[*])	238.36(83.92)
Age of sheep in months	12(11.431)
Male sheep (%)	58.5
Sheep coat color (%)	
Red	44.1
Creamy white	28.7
Black	10.3
Brown	3.1
White	4.6
White mixed with other colors	9.2
Body condition (%)	
Good	48.2
Medium	48.2
Bad	3.6
Body size (%)	
Small	32.8
Medium	42.6
Large	24.6
Tail condition (%)	
Long fat tail	24.1
Long thin tail	47.7
Medium length thin tail	14.9
Medium length fat tail	13.3
Buyers' occupation (%)	
Trader	41
Farmer	10.8
Farmer-trader	6.2
Others	42.1
Sellers' occupation (%)	
Trader	12.8
Farmer	61
Farmer-trader	13.8
Others	12.4

*ETB stands for Ethiopian Birr which is the Ethiopian currency.

anticipated by the buyers and this utility is derived from the attributes of the product as sheep can be considered as quality (attribute) differentiated goods (Lancaster, 1966; Rosen, 1974; Ekeland et al. 2004; Nesheim, 2006). This research focuses on the main phenotypic attributes that buyers inspect when buying a sheep. The external features farmers look at and attach value to are age, fur color, body size, and tail type. The different levels of the homogenous attributes that differentiate sheep are known to both buyers and sellers. The levels considered in this analysis are those perceived by the buyers, despite the possibility of imperfect knowledge and differences in measurement. The buyers and sellers in the markets considered are mainly farmers who raise the sheep. In line with the household modeling literature, where goods are produced, consumed and sold by households, a hedonic model can be employed to value the attributes of the quality differentiated indivisible goods. Therefore, estimation of the relationship between the characteristics of the sheep and their prices can be made through hedonic price modeling.

Following Rosen (1974) and Palmquist (2006), let x_{0j} be the total amount of the jth product characteristic provided to the consumer by consumption of all products, x_{ij} be the quantity of the jth characteristic provided by one unit of product i, and q_i be quantity of the ith product consumed. Then, the total consumption of each characteristic can be given as

$$x_{0j} = f_i\left(q_1, ..., q_n; x_{1j}, ..., x_{nj}\right) \tag{1}$$

and the consumer's utility function is expressed as

$$U = \left(q_1, ..q_n; x_{11}, x_{12}, .., x_{1m}, x_{21}, x_{22}, .., x_{nm}\right) \tag{2}$$

where n is the number of products and m is the number of characteristics.

The consumer is assumed to maximize this utility function subject to a budget constraint that can be specified as

$$Y = \sum_i p_i q_i \tag{3}$$

where Y is fixed money income, and p_i is fixed price paid for the ith product. The consumer's utility maximizing level quantity of each product can then be estimated by maximizing the Lagrangian:

$$L = U(x_{01}, .., x_{0n}) - \lambda\left(\sum_i p_i q_i - Y\right) \tag{4}$$

where λ is the Lagrangian multiplier.

Assuming an interior solution, the first-order condition of the Lagrangian for q_i is given as

$$\frac{\partial L}{\partial q_i} = 0 = \sum\left(\frac{\partial U}{\partial x_{0j}}\right)\left(\frac{\partial x_{0j}}{\partial q_i}\right) - \lambda p_i. \tag{5}$$

It can easily be shown that λ is equal to the marginal utility of income $(\partial U/\partial Y)$. Substituting $\partial U/\partial Y$ for λ and solving for p_i, equation (5) can be rewritten in order to express the demand for attributes as a function of the marginal utility of the attribute and the marginal utility of income.

$$p_i = \sum\left(\frac{\partial x_{0j}}{\partial q_i}\right)\left[\frac{\left(\frac{\partial U}{\partial x_{0j}}\right)}{\left(\frac{\partial U}{\partial Y}\right)}\right]. \tag{6}$$

As income is defined to be equal to expenditure (equation 4), the term in the square bracket is the marginal rate of substitution between expenditure and the jth product characteristics.

Under competitive market conditions, implicit prices will normally be related to product attributes alone, without accounting for producer or supplier attributes. However, as widely documented in the literature, rural markets in developing countries, particularly in sub-Saharan Africa, are rarely competitive (Barret and Mutambatsere 2007).

This is essentially due to poor communication and transport infrastructure, limited rule of law, and restricted access to commercial finance, all of which make markets function much less effectively. Several empirical studies have shown that prices are also related to the attributes of buyers, season and market location (e.g., Oczkowski, 1994; Abdulai, 2000; Jabbar and Diedhiou, 2003). Hence, essential characteristics of the buyer and sellers were included in the models estimated in this research.

Another important issue in estimating hedonic functions is the identification of the appropriate functional form and estimation procedure (Ekeland et al. 2004; Nesheim, 2006). In general, the functional form of the hedonic price equation is unknown (Haab and McConnel, 2002). Parametric, semi-parametric and non-parametric estimations procedures have all been suggested and used in different applications (e.g., Anglin and Gencay, 1996; Parmeter et al. 2007). This research focuses on the estimation of the relative weights of sheep attributes (first step hedonic analysis) and hence the technical details of these alternative approaches are not discussed.

The estimation strategy adopted in this study is a simple linear model based following the suggestion by Cropper et al. (1988) as well as Haab and McConnel (2002). Cropper et al. (1988) employed Monte-Carlo simulation analysis to show that the linear and linear-quadratic functions give the smallest mean square error of the true marginal value of attributes. However, when some of the regressors are measured with error or if a proxy variable is used, then the linear function gives the most accurate estimate of the marginal attribute prices. Haab and McConnel (2002) also argued that when choosing a functional form and the set of explanatory variables, the researcher must bear in mind the almost inevitable conflict with collinearity. High collinearity makes the choice of a flexible functional form less attractive, since the interactive terms of a flexible functional form result in greater collinearity. Given these considerations, we begin with semi-log model given by

$$\ln(\textbf{\textit{price}}) = \textbf{\textit{X}}\textbf{\textit{β}} + \textbf{\textit{ε}}. \tag{7}$$

Following Champ et al. (2003), the market premium (Γ) for an attribute j is computed as:

$$\Gamma_j = 100^* \left(e^\beta - 1\right). \tag{8}$$

In equation (7) the error term is assumed to have a constant variance, σ^2; hence, homoscedastic. However, if and when the errors are heteroscedastic, the OLS estimator remains unbiased, but becomes inefficient. More importantly, the usual procedures for hypothesis testing are no longer appropriate. Given that heteroscedasticity is common in small sample cross-sectional data, methods that correct for heteroscedasticity are essential for prudent data analysis (Long and Ervin, 2000).

Using heteroscedasticity consistent (HC) standard errors is the recommended approach (MacKinnon and White, 1985; Long and Ervin, 2000) to correct for heteroscedaticity of unknown form. The suggested alternative ways of correction using HC include HC_0, HC_1, HC_2, and HC_3. These alternatives are not equally powerful and perform differently under different conditions depending mainly on sample size. Based on Monte Carlo simulation, MacKinnon and White (1985), for example, recommended that in small samples one should use HC_3. However, Davidson and MacKinnon (1993) later recommended strongly that HC_2 or HC_3 should be used. Long and Ervin (2000), similarly, recommended for $N \leq 250$, tests based on HC_2 and HC_3 than those based on other HC. This Monte Carlo simulation result also shows HC_3 is superior for tests of coefficients that are most affected by heteroscedasticity and HC_2 is better for tests of coefficients that are least affected by heteroscedasticity. Accordingly, we have employed HC_2 and HC_3 in this study. OLS was also applied for comparison.

Following Davidson and MacKinnon (1993), the alternative covariance matrix estimators of the error term for HC_2 and HC_3, including the OLS, are specified as:

$$OLS = \frac{\sum e_i^2}{n-k} \left(X'X\right)^{-1} \tag{9}$$

$$HC_2 = \left(X'X\right)^{-1} X' \, diag\left[\frac{e_i^2}{1-h_{ii}}\right] X\left(X'X\right)^{-1} \tag{10}$$

$$HC_3 = \left(X'X\right)^{-1} X' \, diag\left[\frac{e_i^2}{(1-h_{ii})^2}\right] X\left(X'X\right)^{-1} \tag{11}$$

where n is number of observations, k number of parameters estimated, and h_{ii} is x_i' $(X'X)^{-1} x_i$.

Results and discussions

General model results

The results of the hedonic price model from both OLS and HC regressions are given in Table 3. The table summarizes the coefficients of the variables used in the model, and the standard errors of OLS and heteroscedasticity consistent (HC_2 and HC_3) estimations. HC estimation was used as an adjustment to the OLS model since cross-sectional and small sample price data are intrinsically heteroscedastic. As expected, the OLS standard errors were found to be generally lower than the standard errors of HC_2 and HC_3 for all variables except for some variables in HC_2. However, the standard errors of all explanatory variables in HC_3 were increased and greater than both OLS and HC_2 except for three variables brown, thin long tail condition, and sex. Hence, the t-values of the OLS coefficients are inflated and could not be reliable for inferences. Between HC_2 and HC_3, the standard errors in HC_2 were found to be lower than that of HC_3. Therefore, the t-values based on standard errors generated by HC_3 estimation were used for inferences.

Due to the changes in standard errors in the three regression results, significant variables in OLS become insignificant and the significance levels of the variables have also been changed in HC_2 and HC_3. Age square was significant at 5% in OLS but the significance level in HC_2 and HC_3 changed to 10%. Similarly, farmer trader (one of the buyer types) was significant at 5% in OLS, but in HC_2 and HC_3 the variable was significant only at 10%. White mixed coat color, was significant at 1% in OLS but it became significant only at 5% in HC_2 and HC_3. Further, the variable representing trader buyer was significant in OLS and HC_2 but became insignificant in HC_3 estimation.

A model specification test was carried out for the OLS regression model using the Ramsey RESET test. The test with the (null) hypothesis that the model has no omitted variables generated an $F(3, 161)$ value of 0.64 which is extremely below the critical value of 2.65 at $\alpha = 0.05$ implying non-rejection of the hypothesis that there are no

Table 3 Estimation results of OLS and Heterosecdasticity consistent hedonic model

ln(price)	Coefficient	OLS SE	HC2 SE	HC3 SE
Constant	5.2350[‡]	0.0729	0.0725	0.0812
Age and sex				
Age	0.0208[‡]	0.0068	0.0067	0.0076
Age square	−0.0003*	0.0001	0.0001	0.0002
Sex	−0.0457	0.0386	0.0346	0.0372
Coat color				
White	−0.1003	0.0664	0.0965	0.1098
Brown	0.0158	0.0829	0.0615	0.0696
Black	−0.1614[‡]	0.047	0.0488	0.0535
White mixed	0.1305[†]	0.0502	0.0515	0.0563
Creamy white	0.0611*	0.0339	0.033	0.0362
Body Size				
Medium	0.034	0.0247	0.0247	0.0266
Large	0.1467[‡]	0.0414	0.0396	0.0429
Tail condition				
Long thin	−0.0831[‡]	0.026	0.0236	0.0257
Medium & thin	−0.1225[‡]	0.0337	0.0342	0.0373
Medium & fat	0.0319	0.0496	0.0406	0.0454
Season				
Season 1	0.0769[‡]	0.0226	0.0223	0.0239
Season 3	−0.0275	0.0236	0.025	0.0268
Market/place				
Finchaa	−0.0068	0.0309	0.0314	0.0336
G/sanbata	0.0152	0.0306	0.0324	0.0344
Harato	−0.0663[†]	0.0286	0.028	0.0303
Type of seller				
Farmer	−0.0041	0.0262	0.0267	0.0293
Farmer trader	0.0116	0.0375	0.0356	0.0391
Others	−0.001	0.0417	0.0472	0.0518
Type of buyer				
Trader	0.0478	0.028	0.0276	0.0302
Farmer	0.021	0.041	0.039	0.0423
Farmer trader	−0.1042*	0.0506	0.0563	0.0631

‡, † and * significant at $\alpha = 0.01$, $\alpha = 0.05$ and $\alpha = 0.1$ respectively, based on HC_3 standard errors. SE = standard error. Number of observations = 195, $R^2 = 0.6887$.

omitted relevant explanatory variables. The R-square value of the models is 0.6887 implying that the model explained about 69% of change in price of sheep in the local markets of Horro-Guduru Wollega zone of Ethiopia.

Determinants of price and premium for indigenous sheep traits

The results of the three estimations (OLS, HC_2 and HC_3) show that sheep price is determined by sheep traits (such as age, color, body size, and tail condition), season, market places, and buyer type. Age significantly and positively influenced price of sheep in the study area. This is in line with the basic feature of the low input sheep production system in the area. That is, under a low input production system sheep need a longer period of time to attain the required body condition and size to command a good price. Age square, however, influenced sheep price negatively implying sheep command a higher price up to a very old age and then the price will fall down as age goes up. Given the weight of the coefficient and the average age of sheep being marketed, it seems however that old is less of an issue.

From the color dummies, black coat color was found to affect sheep price negatively and significantly. Hence, black colored sheep received a price discount of about 15% as compared to red coat colored sheep. The negative premium the black coated sheep received emanates from the fact that people tend to avoid black coated sheep or any other livestock (see e.g., Kassie et al, 2009) for two reasons. First, the tsetse fly – the vector for sleeping sickness (*trypanosomiasis*) – attacks black coated livestock more? than others. Second, red or other lighter colored sheep are preferred to others for purposes of ceremonial slaughtering in the study area. Whitish and creamy white (locally called *dallecha*) coat colors of traded sheep attracted a 14% and 6.3% price premium respectively, compared to red coat color, *ceteris paribus*. Body size was another trait of sheep that significantly affected price of sheep. Intuitively, sheep with a large body size receive higher prices and hence sheep with a large body size were found to fetch about 15.8% higher price premium compared with small sized sheep. Sheep with thin and long tail and thin and medium length tail received 8% and 11.53% less price, respectively, compared to long and fat tailed sheep.

The determinants of sheep price other than traits of sheep were market location and seasonal factors. Sheep command a significantly higher price in season one (Christmas season) compared with season two (normal season). The Christmas season is the period of high demand for sheep (or livestock in general) that overlaps the crop produce harvesting season that might increase farmers' (as sheep sellers) bargaining power as they can postpone selling when prices are not right. In season one sheep will attract an 8% higher price premium compared to selling in season two. This a general tendency in Ethiopian livestock - particularly small ruminants - marketing as sheep or goat slaughtering is an indispensable part of big festivities such as Christmas provided they are affordable.

Among the market location dummies, sheep in *Harato* attracted a lower price compared with *Shambu*. This is likely due to the relatively high potential for sheep population in the *Harato* area and hence high supply. These results imply that smallholder sheep keepers would benefit if they carefully choose the selling time and the market.

The type of buyer was also an important determinant of price paid for sheep in the study area. Farmer-traders (farmers who do par-time trading) paid a lower price as compared to other groups of buyers. This is possibly because these buyers are well informed both about the production and the marketing of sheep such that they would be in a better position to bargain for a lower price.

In summary, the estimation results show that traits of sheep are much more important determinants of actual price observed than types of buyers and sellers or purposes of buying and selling. Among the attributes considered, age, black coat, large body size, and tail condition were found to be most influential in determining the price paid for sheep in the study area.

Conclusions

This study generated primary data on actual transactions accomplished in four rural markets over three months in central Ethiopia. Using the revealed preference analysis framework and hedonic price modeling, the study determined the level of influence of attributes of sheep and features of buyers and sellers in sheep markets on actual prices paid per head of sheep. As heteroscedasticity is common in cross-sectional data and small sample data, alternative estimations, mainly heteroscedasticity consistent formulations, were employed in addition to OLS estimation.

This estimations have shown how intricate are the relationships between price and traits of sheep and trait and trait level identification of sheep consumers in the rural areas of the study areas. Traits such as age, black coat, large body size, and tail condition were found to be most influential in determining the revealed preferences and hence the prices paid in these rural markets.

Factors such as season of marketing, market location, and type of buyer were also found to be highly important in influencing the prices paid for sheep. The significance of season and market place in influencing price paid for sheep as well justifies the need of targeting season and market places so that smallholder sheep keepers could benefit from the required transformation in the sheep production system. Alternatively or even additionally, linking producers to urban markets where there is high demand for sheep would be an important step to improve farmers' return from the system.

Two important implications can be drawn from the results of this study. First, the consumption of sheep in these areas seem to be very sophisticated such that an intervention that focuses on a single attribute would hardly be successful to improve both supply and demand sides of sheep marketing in the study area. Second, the sheep genetic resources in areas similar to the study areas need to be comprehensively profiled for their attributes. This would be essential in identifying the important attributes of the existing stock and hence prioritizing those traits that need to be improved both for biological and economic efficiency.

Competing interests
The authors declare that they have no competing interests.

Authors' contribution
ZG carried out data collection and supervised enumerators, participated in developing survey instrument, participated in statistical analysis and interpretation of results, participated in designing of the study, and drafted the manuscript. GT participated designing the study, helped in survey instrument development, participated in statistical analysis, and helped in finalizing the draft manuscript. AH participated in designing of the study, helped in preparing survey instrument mainly on identifying sheep trait and categorising these traits as it could be meaningful to researchers in sheep breeding and understandable to respondents. He also helped in editing interpretation of results. DB participated in data analysis and interpretation, and helped in shaping the draft paper.

Acknowledgements
The authors are grateful to ILRI-ICARDA-BOKU project for funding this research. This paper has benefited immensely from insightful comments of an anonymous reviewers and editor of the Agricultural and Food Economics and the authors are very grateful for comments and suggestions received. Yet, the authors alone are responsible for the contents of the paper.

Author details
[1]The University of Liverpool, Liverpool, UK. [2]ICARDA, Aleppo, Syria. [3]ILRI, Nairobi, Kenya. [4]CIMMYT, Harare, Zimbabwe.

References

Abdulai A (2000) Spatial Price Transmission and Asymmetry in the Ghanaian Maize Market. Journal of Development Economics 63:327–349

Anglin PM, Gencay R (1996) Semiparametric Estimation of a Hedonic Price Function. Journal of Applied Econometrics 11:633–648

Barret CB, Chabari F, Bailey D, Little P, Coppock D (2003) Livestock Pricing in the Northern Kenyan Rangelands. Journal of African Economies 12(2):127–155

Barret C, Mutambatsere E (2007) Agricultural Markets in Developing Countries. Entry. In: Durlauf SN, Blume LE (eds) The New Palgrave Dictionary of Economics, 2nd edn. Palgrave McMillan, London

Champ PA, Boyle KJ, Brown TC (eds) (2003) A Primer on Nonmarket Valuation. Kluwer Academic Press, Dordrecht

Chang J, Lusk J, Norwood F (2010) The Price of Happy Hens: A Hedonic Analysis of Retail Egg Prices. Journal of Agricultural and Resource Economics 35(3):406–423

Cropper ML, Deck LB, McConnel KE (1988) On the Choice of Functional Form for Hedonic Price Functions. Rev Econ Stat 70(4):668–675

CSA (Central Statistical Agency) (2009) Statistical Abstract. Addis Ababa, Ethiopia

CSA (Central Statistical Agency), (2007) Statistical Abstract. Addis Ababa, Ethiopia

Davidson R, MacKinnon JG (1993) Estimation and Inference in Econometrics. Oxford University Press, Cambridge

Drucker AG, Gomez V, Anderson S (2001) The economic valuation of farm animal genetic resources: A survey of available methods. Ecol Econ 36:1–18

Ekeland I, Heckman JJ, Nesheim L (2004) Identification and Estimation of Hedonic Models. Journal of political economy 112(1): s60–s109

FAOSTAT (2009) FAO (Food and Agricultural Organization of the United Nations), Rome, http://faostat.fao.org. Accessed 19 October

FAOSTAT (2010) FAO (Food and Agricultural Organization of the United Nations), Rome, http://faostat.fao.org. Accessed 23April

Faustin V, Adegbidi AA, Garnett ST, Koudande DO, Agbo V, Zander KK (2010) Peace, health or fortune?:Preferences for chicken traits in rural Benin. Ecol Econ 69(9):1848–1857

Gezahegn A, Mohammad AJ, Hailemariom T, Elias M, Getahun K (2006) Seasonal and Inter-Market Differences in Prices of Small Ruminants in Ethiopia. Journal of Food Products Marketing 12(4):59–77

Haab TC, McConnel KE (2002) Valuing Environmental and Natural Resources: The Econometrics of Non-Market Valuation. Edward Elgar, Cheltenham, UK, Northampton, MA, USA

Hensher D, Rose J, Greene W (2005) Applied Choice Analysis: A Primer. Cambridge University Press, Cambridge

Jabbar MA, Diedhiou ML (2003) Does Breed Matter to Cattle Farmers and Buyers? Evidence from West Africa? Ecol Econ 45:461–472

Kassie GT (2007) Economic Valuation of the Preferred Traits of Indigenous Cattle in Ethiopia. Christian Albrechts University of Kiel, Kiel, Germany, PhD Dissertation. pp, 177

Kassie GT, Abdulai A, Wollny C (2009) Valuing Traits of Indigenous Cows in Central Ethiopia. Journal of Agricultural Economics 60(2):386–401

Kassie GT, Abdulai A, Wollny C (2011a) Heteroscedastic Hedonic Price Model for Cattle in the Rural Markets of Central Ethiopia. Applied Economics 43(24):3459–3464

Kassie GT, Abdulai A, Wollny C, Ayalew W, Dessie T, Tibbo M, Haile A, Mwai O (2011b) Implicit prices of indigenous cattle traits in central Ethiopia: Application of revealed and stated preference approaches. ILRI Research Report 26. CIMMYT and ILRI, University of Kiel

Lancaster K (1966) A new approach to consumer theory. Journal of Political Economy 74:132–157

Long JS, Ervin LH (2000) Using Heteroscedasticity Consistent Standard Errors in the Linear Regression Model. The American Statistician 54(3):217–224

MacKinnon JG, White H (1985) Some Heteroscedasticity Consistent Covariance Matrix Estimators with Improved Finite Sample Properties. Journal of Econometrics 29(3):305–325

Negassa A, Jabbar M (2008) Livestock ownership, commercial off-take rates and their determinants in Ethiopia. Research Report 9. ILRI (International Livestock Research Institute), Nairobi, Kenya, p 52

Nesheim L (2006) Hedonic price functions. Centre for microdata methods and practice working paper CWP18/06

Oczkowski E (1994) A Hedonic Price Function for Australian Premium Table Wine. Australian Journal of Agricultural Economics 38:93–110

Omondi I, Baltenweck I, Drucker AG, Obare G, Zander KK (2008) Economic valuation of sheep genetic resources: implications for sustainable utilization in the Kenyan semi-arid tropics. Trop Anim Health Prod 40(8):615–626

Palmquist RB (2006) Property Value Model. In: Maler KG, Vincent J (eds) Handbook of Environmental Economics, Edition 1, 2nd edn., pp 763–819

Parmeter CF, Henderson DJ, Kumbhakar SC (2007) Nonparametric Estimation of a Hedonic Price Function. Journal of Applied Econometrics 22:695–699

Richards TJ, Jeffrey SR (1996) Establishing indices of genetic merit using hedonic pricing: an application to dairy bulls in Alberta. Canadian journal of Agricultural Economics 44:251–264

Rosen S (1974) Hedonic Prices and Implicit Markets: Product Differentiation in Pure Competition. Journal of Political Economy 82(1):34–35

Satimanon T, Weatherspoon D (2010) Hedonic Analysis of Sustainable Food Products. International Food and Agribusiness Management Review 13(4):57–74

Scarpa R, Kristjanson P, Ruto E, Radeny M, Rege JEO (2003) Valuing indigenous cattle breeds in Kenya: an empirical comparison of stated and revealed preference value estimates. Ecol Econ 45:409–426

Tibbo M (2006) Productivity and health of indigenous sheep breeds and crossbreds in the Central Ethiopian highlands. PhD Thesis, Swedish University of Agricultural Sciences. Sweden University Press, Uppsala

Williams OT, Okike I, Spycher B (2006) A Hedonic Analysis of Cattle Prices in the Central Corridor of West Africa: Implications for Production and Marketing Decisions. In: Contributed paper prepared for presentation at the International Association of Agricultural Economists Conference. Gold Coast, Australia

3

Institutional reforms and agricultural policy process: lessons from Democratic Republic of Congo

Catherine Ragasa[1*†], Suresh C Babu[2†] and John Ulimwengu[3†]

* Correspondence: c.ragasa@cgiar.org
†Equal contributors
[1]Development Strategy and Governance Division, International Food Policy Research Institute (IFPRI), 2033 K Street, NW, Washington DC 20006, USA
Full list of author information is available at the end of the article

Abstract

Attaining food security for all requires well-functioning institutions and policy process that are effective in designing and implementing food and agricultural policies and programs. This paper assesses early stages of the decentralization and institutional reforms in the policymaking processes in the Democratic Republic of Congo (DRC). It develops a conceptual framework and adopts an innovative mapping tool to identify capacity and incentive challenges impeding the effective design and implementation of food and agricultural policy and institutional reform processes. We found that decentralized platforms for policy dialogues have the potential to improve civil society participation in local-level and national-level policy and planning processes. However, their success depends on organizational and human capacity strengthening, financial sustainability, effective participation of the civil society, and demonstrated impact of their participation.

JEL Classification: D72; D74; D78; Q18

Keywords: Food security; Institutional reforms; Policy process; Decentralization; Network mapping; Capacity development; Post-conflict; Agriculture

Background

Attaining food security for all in countries that are behind in achieving Millennium Development Goals requires institutional and policy reforms that will help the food and agricultural system to become more productive and efficient in the use of natural resources. The role of institutions, governance, and the capacity of countries in efficient use of development assistance remain on the top of the research agenda of the development community (Killick 2010). However, how such research can result in better development practice such as removal of barriers to development investments is still to be seen (Booth 2011). Institutional reforms and innovations require recognition of the nature and state of organizational and human capacity[a] in the country context and how such capacity can be further developed to achieve clear development goals (Grindle 2007). The challenge for the development organizations is to understand the process of institutional reform in order to facilitate positive changes in the specific practical, economic, and organizational contexts. Learning from the institutional reform experiences and the policy processes[b] is vital for increasing effectiveness of development assistance. In this paper we use the case of Democratic Republic of Congo

(DRC) and focus on its policymaking and capacity development processes to derive specific lessons for the countries facing challenges in achieving food security for all.

In post-conflict societies, such as DRC, opportunities arise for reforms in policies and institutions. Yet, serious capacity gaps[c] often constrain the effective design and implementation of institutional reforms. DRC, after episodes of war that devastated much of its human capital and physical infrastructure, provides a good setting to analyze the opportunities for policy and institutional reforms as well as sequencing and complementing these reform processes with capacity development approaches. DRC is considered as a key agricultural producer in Africa before the war; and estimates suggest that if agricultural productivity were to catch up with the global technological frontier, DRC could feed around one-third of the world's population (Tollens 2004). It is endowed with 80 million hectares of arable land (of which 4 million are irrigable), 125 million of tropical forest representing 6 percent of world forest reserves, climatic diversity allowing multiple agricultural seasons in the same year, grazing land capable of supporting nearly 40 million head of cattle, inland fisheries resources that can enable annual production of at least 700,000 tons of fish, and rich ecology and abundant hydrology that can allow the practice of various farming activities (World Bank 2006).

In an effort to increase economic development by rebuilding its agriculture and economic sectors, DRC is embarking on several institutional reform processes, including more demand-driven, inclusive, and evidence-based policymaking and planning. In the agricultural sector, the country has set up Agricultural and Rural Management Councils (CARGs), which are a platform for discussion, information sharing, and formulation of local agricultural strategies, in its territories and provinces. Additionally, DRC is restructuring and decentralizing its Ministry of Agriculture, Fisheries, and Livestock (MINAGRI) and Ministry of Rural Development (MINRD) to make them efficient and responsive to the needs of the population. In June 2010, DRC committed to increase its agricultural GDP and its budget share for agriculture through the launch of the Comprehensive Africa Agricultural Development Programme (CAADP) framework. A number of international organizations have supported the agricultural sector through numerous projects and programs that provide basic and economic services directly to beneficiaries to help them increase their income. However, none of them has looked at the broader organizational capacity constraints involved in policymaking and planning. For increasing the effectiveness of aid, it is important to identify such organizational and capacity constraints and to understand how donor support could "build on" what is already there, in order to remove disincentives and blockages in the process of evidence-based policy making (Grindle 2007). An innovative methodology is used to study the policy process in the DRC agricultural sector in the context of the processed institutional reforms and organizational and capacity constraints.

This paper provides two unique contributions to the literature. First, it offers a conceptual framework of analyzing the nexus of incentives and capacity in institutional and policy reform processes. While the incentives and civil-service reforms have been recognized in the literature, they have rarely been included or paid attention to in training needs assessment or capacity assessment and subsequent training and capacity development activities. Second, this paper applies a simple, yet powerful, tool for rapid assessment to identify key actors, their linkages, constraints, capacity, and incentives. In addition, this paper aims to inform policymakers, stakeholders, and international partners on the constraints and feasible options available to more fully develop agricultural sector.

The paper is organized as follows. A conceptual framework is developed in the next section after a brief review of institutional reform and policy process literature. In section three, we present the methods used. In section four, we apply the framework and methods to study the recent institutional reform processes in the food and agricultural sector of DRC. Section 5 draws specific lessons and implications from the case study. conclusion forms the last section.

Literature review and a conceptual framework

Two strands of literature guide the process of developing a conceptual framework to study institutional reforms in the context of capacity constraints. First, the policy-processe literature helps to identify the institutional reforms that needed for evidence-based policymaking in developing countries; and second, the organizational and capacity development literature helps to understand various factors affecting the organizational performance of institutional reform process. In addition, recent scholarship places high emphasis on understanding institutional and policy responses during emerging challenges such as food crisis and resilience of developing countries to face such crisis (Babu 2013).

While several theories exist, empirical investigation of policy process in the context of developing countries is scarce (Court and Young 2003; Omamo 2004; Sabatier 2007; Resnick and Birner 2010). Research–policy linkages, a subset of the policy process literature, have received some attention (see Ryan 1999; Guston 2001; Stone et al. 2001; Court and Young 2003; Cash et al. 2003; Young 2005; Ayuk and Marouani 2007; and Aberman et al. 2010). Ayuk and Marouani (2007) assess multiple case studies of policy–research links and highlight the nonlinear nature of the link between research and policy, which requires flexibility and agility on the part of researchers to seize opportunities and to quickly reassess and recalibrate research approaches as needed. Aberman et al. (2010) find that champions of research within policymaking circles are key to the application of research to policy, and that strong ties between researchers and technocrats likewise played an essential role in ensuring government buy-in. Despite this body of literature on policy processes and research, obvious gaps remain.

Capacity development process and policy process are widely discussed in the literature as two separate disciplines; they are rarely analyzed systematically together. For example, on one hand, a major stumbling block for decentralization and privatization policies in extension services delivery is the lack of proper capacity assessment and subsequent investments in capacity development; while major constraints in pluralistic and demand-driven service delivery systems are the lack of capacity for administration and management of funds as well as a shortage of service providers (see Rivera and Alex 2004; World Bank 2010). On the other hand, training and capacity development efforts are not effective and will not lead to desirable impacts without enabling policy environment, and institutional and organizational reforms (Adebayo et al. 2009; Babu et al. 2007), and civil service reforms (World Bank 1999). How these two processes are interlinked, how to design and implement interventions that contribute to the two processes, and how they are effectively sequenced remain a major research gap.

Given this backdrop, this paper addresses the current research gap in understanding the role of institutional reforms in the evidence-based policy process. It develops a framework and applies an expanded mapping tool that incorporates elements of capacity and incentive conditions surrounding policy processes.

The framework centers on a clearly defined development goal (or outcome) of interest that the country or sector is aiming to achieve (Figure 1). Agricultural development can be achieved by aspiring for efficiency, effectiveness, and sustainability of organizations that are effective in key processes in the agricultural sector management, namely: (1) inclusive and evidence-based policymaking and planning, (2) adequate and predictable resources, (3) effective and demand-driven service provision, and (4) enforceable regulations. Outcomes of these processes will have to be combined to produce development impacts. These collective functions of agricultural-sector management have to be guided by the strategic direction and measurable performance targets broadly shared among the actors and organizations.

The organizational performance and effectiveness of key processes are in turn conditioned by incentive and capacity, which are influenced by the organizational or institutional landscape, country context, and broader enabling environment. *Organizational capacity* is characterized by management systems and procedures for coordination and communication, and availability and adequacy of financial and physical resources. *Human capacity*, which reflects the quantity and quality of human resources and social capital, is the summation of skills, knowledge, and competencies of individuals. It is also a crucial factor in achieving pre-defined objectives and development goals. Education and training systems of a country play a critical role in capacity development; while well-designed external support to the education system or directly to the organizations is potentially important in moving out of path dependency and embracing reforms and change.

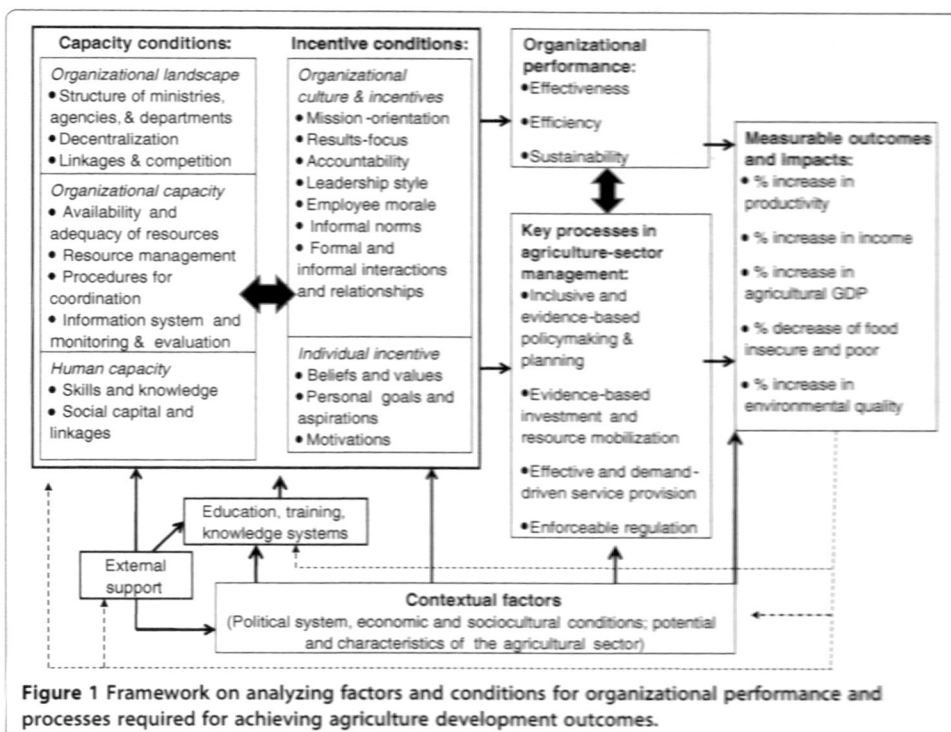

Figure 1 Framework on analyzing factors and conditions for organizational performance and processes required for achieving agriculture development outcomes.

Incentives are inherent to individuals' preferences and needs and influenced by the nature of both formal and informal institutions or norms at different levels (country, sector, organization, or unit level) and can be manifested in the degree of external checks and balances and organizational culture. *External checks and balances* in the form of client feedback and other related external pressures are likely to elicit demand-side accountability and are capable of influencing behavior if grounded with a credible incentive system to perform. Incentives can be in the form of monetary or nonmonetary benefits of a particular decision or action. *Organizational culture*, which represents "the collection of traditions, values, policies, beliefs and attitudes that constitute a pervasive context for everything we do and think in an organization" is an important factor that may affect incentives of individuals to perform (Marshall and McLean 1988, page 32).

In the next sections we apply this conceptual framework to study the organizational and human capacity constraints and challenges as they relate to the institutional reforms undertaken in the DRC agricultural sector.

Methods

Qualitative assessment through mapping tools is used for this case study to provide a rapid assessment of the capacity and incentive constraints faced by the actors and to suggest possible entry points for support used to analyze governance linkages, social networks, and implementation processes of programs and projects. Process-network-influence mapping (or simply Net-Map) is an interview-based mapping tool that helps users to understand, visualize, discuss, and improve situations in which many different actors influence outcomes (Schiffer 2007). This paper extends this by adding a series of interview questions about constraints faced by key actors and identifies whether these constraints are due to lack of capacity or lack of incentive.

A literature review on key organizations involved in the agricultural sector in DRC provided a tentative list of key officials or representatives who were contacted for interviews. The team started institutional mapping exercise with a key official at MINAGRI, which helped guide a snowball approach to selecting the subsequent interviewees. The study team conducted 26 interviews (with 46 key experts) in Kinshasa between May and September 2010. A total of 7 Net-Map interviews (with individuals or groups) and 19 semi-structured interviews (with individuals or groups) were conducted.

The Net-Map tool starts with a well-designed research question, expressed in the form of a final outcome or development objective. The focus is on the actors and their linkages, influence level, and capacity and incentive constraints in the actual recommendations and implementation of policy priorities stated in the Agricultural Code, Agricultural Policy Note, and Agricultural Strategy of DRC. The first phase of the mapping identified the key actors. Respondents described the policy processes step-by-step (from evidence-based research to actual funding and adoption of the policy) and identified the actors involved in each step.

The second phase involved mapping the linkages and interaction among the different actors. In communicating the definition to interviewees, the interviewers stressed that interaction or linkages can be in the form of funding, information flow, research, advice, advocacy, and reporting/authority and also asked respondents to describe the form of linkages they formed with other stakeholders. Those interviewed were asked to rate the level of this interaction using a 6-point Likert scale (1 = *no linkage*, 2 = *very weak linkages*, 3 = *weak linkages*, 4 = *somewhat strong linkages*, 5 = *strong linkages*, 6 = *very strong linkages*).

The third phase of the mapping involved asking respondents to rate the influence that different actors have on the actual policy process including recommendations and implementation in the Agricultural Code, Agricultural Policy Note, and Agricultural Strategy of DRC. To determine influence, the interviewers stressed that all types of influence are considered—financial, formal influence, communication, advice, or voice—in determining the final influence level.

The rating was done on a scale from 0 to 6 and was visualized by using Checkers game pieces. As shown in Figure 2, the actors are indicated by the board game figures, and the Checkers game pieces are used to build towers, the height of which shows the influence level of the respective actors. This method has the advantage of visualizing influence levels in a three-dimensional space. While performing this exercise, the respondents were also asked to identify why different actors have the influence level ascribed to them.

For human capacity issues, respondents were asked to identify the strategic capacity-strengthening activity that would make the greatest positive impact in their organization. This emphasis during the interviews provided useful insights into the urgency and prioritization of investments in training needed to support the agricultural sector. Skill gaps and training needs had been identified by interviewees' self-evaluation.

All the results of the maps were consolidated into a single map. To capture the nuanced information attained through the interviews, the study team summarized the different maps drawn and the qualitative information gathered in the key informant interviews, taking into account the frequency with which actors and links were added and influences were attributed, the extent of involvement in the policy process of the respondents who added them, and the goals and possible biases of the particular respondents that may have affected their answers.

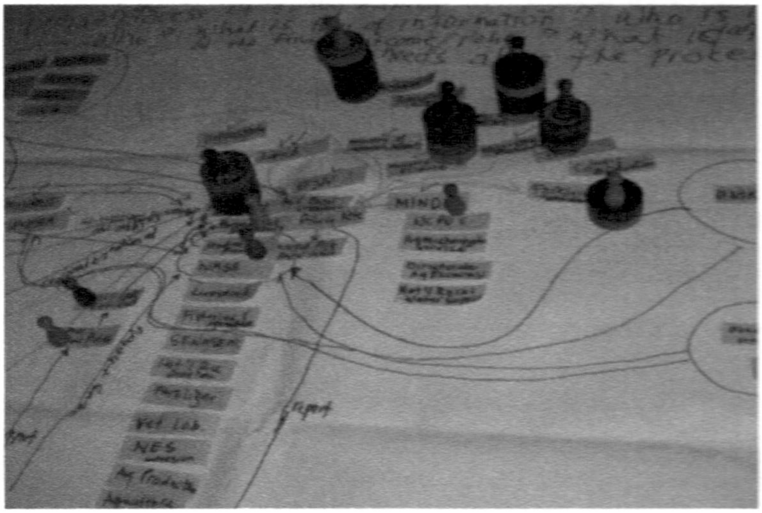

Figure 2 Net-Map of influence and power relation among key actors.

The development of the final map can be described as an iterative process in which the study team collected the preliminary perception and information from existing literature, gathered information from a wide variety of actors involved in the policy process, drew a combined network map based on that information, and finally validated it by presenting and discussing it with key officials and experts in MINAGRI and other interviewees. The resulting map serves as a visual representation of the policy process and is used to illustrate the core actors, linkages, influence level, bottlenecks, incentives, and capacity constraints and for evidence-based and inclusive policy processes.

Results and discussion

Using the conceptual framework described in section 2, this section presents the results of the analysis of the data collected through the methods presented above.

Organizational landscape

Figure 3 presents the organizational landscape of the agricultural sector in DRC based on interviews with 45 high-level officials and representatives of the MINAGRI, the Ministry of Rural Development (MINRD[d]), CARGs, parliament, development NGOs, the private sector, and universities.

Table 1 presents a summary of the roles of the different key organizations in the agricultural policy process. MINAGRI and MINRD are the two main ministries responsible for leading the agricultural strategy formulation and policymaking process in DRC. The

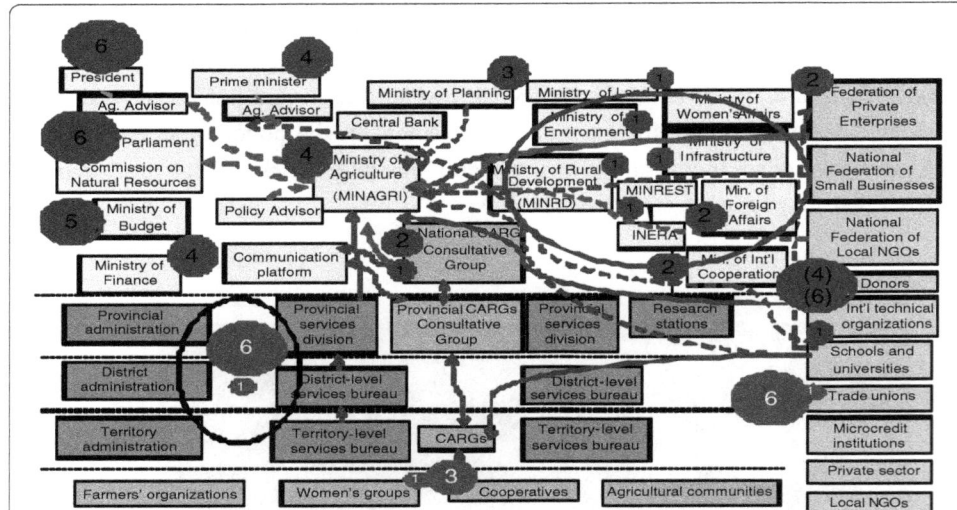

Figure 3 Organizational landscape, linkages, and influence level of key actors in the agriculture policymaking process in DRC. Net-Map interviews with a high-level official of the ministry of agriculture, CARG representatives, and a staff member of an international organization who has worked in DRC for several years. Note: These Net-Maps were complemented by interviews with several key actors in the agricultural sector. Note: The structure and placement of the actors in the figure do not represent political ranking or authority. The purpose is to inventory the different actors involved in the agricultural policy process at the national and local levels; to illustrate linkages among different actors; and to identify who has influence in the actual agricultural policies being designed, funded, and implemented. The numbers in circles represent the influence level (measured on a 0–6 scale) of these different actors, with darker circles for the national-level actors and lighter circles for the local level. The lines connecting key actors represent the extent of linkages and information flow between the actors (solid lines represent strong linkages while broken lines represent weak linkages). Organizations without an influence score were given a zero score by respondents.

Table 1 Roles and influence level of different actors in the agricultural policymaking process in DRC

Key organizations/individuals	Roles in policy process and source of influence	Influence level
President, parliament	They lead the national priorities, which in turn influence the budget allocation and disbursement to the different sectoral ministries and public services.	6
Donors	They provide support, through funding, advice, analytical work, technical assistance, and capacity-strengthening activities, among others. Due to limited local resource mobilization, the government relies heavily on donor funding. Given the lack of analytical capacity, the government relies on analytical work and advice from donors. The influence score assigned to donors vary by interviewees.	4–6
Ministry of Budget	It decides on actual budget allocation for the sector ministries.	5
Ministry of Finance	It sets priorities on actual budget disbursement.	4
MINAGRI	It coordinates the consultative process, design, drafting, implementation, and monitoring of policies, strategies, and plans for the agricultural sector; it leads the resource and political support mobilization.	4
Prime Minister	Together with the president, the prime minister leads the country and oversees national priority setting.	3
Ministry Planning	It sets the national economic plan that affects agricultural strategies, programs, and budget.	3
Federation of Private Enterprises (Federation des Enterprises du Congo, or FEC)	This serves as a forum for dialogue among larger private-sector firms and a platform for advocating their concerns, needs, and priorities to MINAGRI in particular and policy processes in general.	2
CARG	It is a platform for policy dialogues and discussions, sharing problems and jointly finding solutions, setting priorities and strategies at the local level, and advocating for local demands and needs at the national policy level.	1–2
MINRD, Ministry of Land, Ministry of Infrastructure, Ministry of Research, and others	MINRD is responsible for community development, rural roads/feeder roads, small rivers, rural water supply, and rural housing. All of these ministries provide inputs to policy design and lead the implementation of activities and programs specific to the ministries' mandates.	1
International technical organizations (FAO, IFPRI, other CGIAR centers)	They provide technical support to the many functions of MINAGRI and other ministries.	1
Ministry of International Cooperation, Ministry of Foreign Affairs	They facilitate and manage the international partnerships in DRC.	1
Universities (UNIKIN, University of Lubumbashi, University of Yangambi, and UCC)	They are tasked to provide research and technical advice on policy options. However, they are not currently linked to MINAGRI and other ministries or policy processes.	0
Agricultural education and training institutes	They are tasked to train educators, researchers, extension agents, and professionals to work for the sector (government, private sector, NGOs, etc.). However, they do not have a voice and direct engagement in the actual policy process.	0
COPEMECO and other national federations	They are tasked to represent the interest of small business, farmers' organizations, and civil society at large. However, they do not have a voice in the actual policy decisions.	0

Source: Author's interviews and Net-Maps.

political system that conditions the financial resources and enabling environment for the agricultural sector includes the President, Prime Minister, and parliament at the national level and elected government administrators at the local level. Political parties play a role in the selection of candidates for elected and appointed positions at local and national levels. They can also be involved in the recruitment and promotion of staff in the public sector.

The government ministries are responsible, each in its area of intervention, for policy, planning, implementation, and monitoring of development programs. In total, there are 11 provinces (including the capital city Kinshasa), 30 districts,[e] 145 territories, and 800 sectors (formerly called *collectivities*) in DRC. Within the decentralization process, the provincial services of the ministries are being revitalized to better serve the rural population. Staff recruitment to support the decentralized units is ongoing.

A number of universities perform some agricultural economic and rural development research, namely, University of Kinshasa (UNIKIN), University of Lubumbashi, University of Yangambi, and the Catholic University of Congo (UCC). International organizations provide technical assistance and analytical support, sometimes in collaboration with these local universities. The main donors for the agricultural sector in DRC include the BTC, World Bank, European Union, and USAID.

Farming communities and their organizations are crucial players in the agricultural sector. The three federations of NGOs—the Federation of Laic and Economic NGOs (FOLECO), CNONGD and the Regional Council of Development NGOs (CRONGD)—have so far not gained a strong influence in the agricultural policymaking processes. For a more systematic engagement of civil society organizations into policymaking and planning processes, the government initiated in 2008 the creation of CARGs, which are the platforms for multi-stakeholder discussions and dialogues toward the development of agricultural plans at the territorial, provincial, and national levels. Private-sector participation in the agricultural policymaking process in DRC has not been strong in the past years, except for some advocacy efforts by the FEC (Federation of Private Enterprises in Congo). Similarly, the Confederation of Small and Medium Enterprises of the Congo (COPEMECO), of which numerous small-scale enterprises are members, have not gained a strong presence in the agricultural policymaking process in the past.

Influence

Understanding the influence level of each organization in the policy and institutional reforms process is important to identify feasible entry points for intervention; to prioritize capacity and incentive gaps; and to identify weaknesses in information flow and links between research-based evidence and decision-making process. As indicated in Figure 3 and Table 1, high-level policy- and lawmakers (the President and the Members of Parliament) were given the highest influence score (6) because they set national priorities, which in turn influence the budget allocation and disbursement to the different sectorial ministries and public services. The next most influential is the Ministry of Budget, which decides on actual budget allocation for the sector ministries, with an influence score of 5, followed by the Ministry of Finance, which sets priorities on actual budget disbursement (influence score = 4). The prime minister also has strong influence in setting national priorities and agricultural policies. Depending on the interviewer, the donors in general were given a score of 4 or 6 in terms of their level of influence.

The MINAGRI is responsible for leading the agricultural strategy formulation and policymaking process in DRC (influence score = 4), while other relevant ministries, such as MINRD, provide inputs and support (collective influence score = 1). FEC was singled out as influential in the actual policies formulated in MINAGRI (influence score = 2), while civil society, through the CARGs, also has only small influence in the actual policy formulation to date (influence score = 1–2). The Ministry of International Cooperation and Ministry of Foreign Affairs, which facilitate and manage the international partnerships in DRC, were given a score of 2 in terms of influence, while international technical organizations (particularly FAO and other UN agencies) were given an influence score of 1. These patterns of influence suggest that greater attention is needed to ensure that compelling evidence from research reaches the influential decision-makers and funders. This will require strengthening policy dialogue and negotiation skills, among many other skills and competencies. It may also require more systematic mechanisms of linking MINAGRI and these decision-makers, such as initiating regular meetings with the agriculture advisors of the president and prime minister, and to examine the re-establishment of the socioeconomic research committee, with emphasis on the agriculture and rural sectors, at the parliament.

At the provincial level, members, coordinators, focal points of CARGs and representatives of MINAGRI were asked to rate the level of influence of the different actors at the provincial and territory levels on the local agricultural policy outcomes (lighter circles in Figure 3). Respondents highlighted that experiences across provinces and territories vary. Within CARGs, in theory the level of influence should be in favor of the civil society as it represents two-thirds of the membership, with the government having one-third. In reality for many places, the most influential (with a score of 6) are still the government and/or ministry officials at provincial and territory levels, while in some areas the most influential (with a score of 6) are trade unions. The level of influence by the government and trade unions can also score as low as 1 depending on the provinces and territories according to key informants. There is some influence by the producer's association and civil society organizations through CARGs, with a score of 1 to 3 depending on the province or territory. Respondents indicate an improvement in the civil society participation compared with previous years, and they are hopeful that it will continue to improve as the implementation of and support for CARGs goes into full speed. The CARG guidelines also indicate that one-third of CARG membership should be women, but either lack of women leaders or inability of women leaders to attend CARG-related meetings and workshops hinders the realization of this quota of participation in CARGs.

The pattern of influence at the local level suggests the need to strengthen civil society organizations so that they would be able to participate more effectively and benefit in platforms and networks, such as CARGs. Capacity strengthening will also be needed by public-sector organizations as they remain major players that lead, guide, and influence CARG functioning. CARGs depend on the capacity and strength of the members organizations. The Net-Map exercise suggests that experiences differ by provinces and territories, so further studies will be needed to understand influence level, actual functioning, and impact of CARGs. Further analysis will also be needed to study the constraints of women in participating in CARGs and to identify whether this is a constraint in the effective and sustainable functioning of CARGs.

Linkages

Table 2 provides the linkage scores among organizations rated by respondents from different organizations in DRC. Key informants have highlighted a number of weaknesses

Table 2 Degree of linkages among organizations in the agricultural policy process

Linkage	Avg. score*	Std. dev.	Min.	Max.
Between MINAGRI and INERA	4.8	0.9	3.0	6.0
Between MINAGRI and CARG	4.6	1.0	3.0	6.0
Between MINAGRI and international partners	4.5	1.1	2.0	6.0
Between INERA and universities	4.2	1.2	2.0	6.0
Between MINAGRI and MINRD	4.1	1.2	2.0	6.0
Between MINAGRI and Ministry of Planning	4.0	1.2	2.0	6.0
Between MINAGRI and Min. of Finance/Budget	3.9	1.0	3.0	6.0
Among universities	3.9	1.2	2.0	5.0
Between national-level and local-level MINAGRI	4.0	1.3	1.0	6.0
Between national-level and local-level MINRD	3.8	1.1	2.0	6.0
Between universities and international partners	3.8	1.3	2.0	6.0
Between MINAGRI and offices of president, prime minister, and parliament	3.7	1.0	2.0	6.0
Between MINAGRI and NGOs	3.6	1.3	1.0	5.0
Between MINAGRI and universities	3.6	1.2	1.0	5.0
Between MINAGRI and other ministries	3.5	1.3	1.0	5.0
Between CARGs and NGOs at local levels	3.5	0.8	2.0	5.0
Between MINAGRI and private sector	3.4	0.7	2.0	5.0
Between CARGs and CRONGD and CNONGD	3.3	1.2	1.0	5.0
Between CARGs and FEC	3.3	1.2	1.0	5.0
Between CARGs and FOLECO	3.2	1.0	1.0	5.0
Between CARGs and COPEMECO	3.1	1.3	1.0	5.0
Between CARGs and INERA	3.0	1.1	1.0	5.0
Between CARGs and universities	2.8	1.1	1.0	5.0
Between universities and private sector	2.8	0.9	1.0	4.0

Source: Authors' compilation based respondents.
Note: * Interviewees were asked to rate the level of interaction and linkages between various organizations using a 6-point Likert scale (1=no linkage, 2=very weak linkages, 3=weak linkages, 4=somewhat strong linkages, 5=strong linkages, 6=very strong linkages).

in the interaction and linkages among actors in the agricultural landscape (illustrated by the broken lines in Figure 3). Overall, MINAGRI's linkage with other organizations involved in the policy processes in DRC is weak. The major missed opportunity is between policy analysis and the highest levels of influence. As indicated above, the national decision-makers (including the president, parliament, and prime minister) are not reached systematically by sources of evidence and cutting-edge research on the role and importance of the agricultural sector and on viable options and priorities of advancing the sector. Although the offices of the president and prime minister do have agriculture advisors, key informants suggest only weak interaction between them and MINAGRI. There is also a commission in the parliament that looks at agriculture and natural resources policies, but key informants also suggest a weak linkage between members of this commission and MINAGRI. In Table 2, an average linkage score of 3.7 (out of 6), which ranges from weak to somewhat strong linkage, indicates a huge space for improvement to fill the research–policy–action gap. Moreover, the agencies that decide on budget allocations and actual disbursements are not systematically reached by convincing and compelling evidence of investing in the sector as well as by advocacy and lobbying that make the case for increased investment in agriculture (mean linkage score = 3.9). This influence-linkage analysis highlights the mismatch between sources of research and sources of influence, which appears to be a major hurdle in evidence-based planning in DRC.

The two main ministries that lead the agricultural and rural sectors, MINAGRI and MINRD, have some degree of coordination (mean linkage score = 4.1), but respondents indicated that this could be improved by clarifying roles and responsibilities for each ministry and communicating ongoing reform and change processes.

In terms of priority setting and strategic planning, MINAGRI gets technical inputs from the inter-ministerial committee, although with limited interaction (mean linkage score = 3.6). MINAGRI sources its evidence-based analysis from international organizations (e.g., World Bank, FAO, BTC, and IFPRI) and has relatively strong linkages with these partners (mean linkage score = 4.5). Both MINAGRI and MINRD get agricultural data through the statistical and data-collection systems in the provincial and territory services units, but interaction and coordination between national and local units in these ministries remain limited (mean linkage score = 3.8–4.0). They receive feedback on problems and needs of the rural population through the CARGs platform, which are part of MINAGRI (mean score linkage = 4.6); but there is limited interaction between MINAGRI or MINRD and FEC (mean linkage score = 3.4) and the federation of NGOs (mean linkage score = 3.7).

In terms of evidence-based analysis, the capacity at MINAGRI is weak, but the capacity for economic and policy research in other organizations (universities, Central Bank's Economic Research Department, and Ministry of Planning) is relatively stronger (Table 3). The issue is that the linkage between the ministry and these organizations with capacity is weak to somewhat strong (mean linkage scores = 3.6–4.0). Key informants indicate that this limited interaction is due to lack of trust from both sides and lack of incentive to interact as dictated by a different set of performance indicators in these organizations. Currently, there is no capacity at MINAGRI to initiate or facilitate such interaction and linkage. On the other hand, the linkage between MINAGRI and INERA seems to be stronger for scientific research (mean linkage score = 4.8). MINAGRI should use this example to follow in strengthening interactions with other research organizations and universities.

Universities have started to be actively involved in providing results of analytical work through the organized technical workshops within CARGs in relation to the provincial agricultural plan formulation. However, outside this mechanism, there are no other informal or formal linkages between universities and members of the CARGs. While there is stronger interaction within universities (mean linkage score = 3.9) and between universities and INERA (mean linkage score = 4.2), there are weak linkages between these universities and CARGs (mean linkage score = 2.8) and between universities and the private sector (mean linkage score = 2.8). Universities have only limited opportunities for research and consultancy with international partners (mean linkage score = 3.8). Interviews also suggest weak linkages between universities and other ministries and public-sector agencies. Moreover, policy advisors of MINAGRI are responsible for providing cutting-edge and up-to-date advice in the ministry, but insights from the interviews suggest that there is limited interaction between policy advisors of the minister and universities and sources of analytical work. Similarly, agricultural policy advisors inform the president and prime minister, but they are currently not linked to the universities, ministry of agriculture, or sources of information and analytical work.

Table 3 Capacity levels of key organizations in the agricultural policy processes

Capacity/Organization	Human capacity		Facilities and infrastructure		Organizational procedures and management systems		Monitoring and evaluation system		Coordination and communication systems	
	Avg. score*	Std. dev.	Avg. score	Std. dev.	Avg. score	Std. dev.	Avg. score	Std. dev.	Avg. score	Std. dev.
Overall	3.8	1.2	3.2	1.2	3.3	1.1	3.0	1.2	3.3	1.1
Main universities										
UNIKIN	4.6	1.1	3.7	1.3	3.6	1.2	3.3	1.3	3.6	1.3
UCC	4.5	1.0	4.3	1.3	4.0	1.1	4.1	1.1	4.0	1.2
U. of Yangambi	3.8	1.1	3.1	1.4	3.3	1.0	3.1	1.2	3.1	1.3
U. of Lubumbashi	4.3	1.2	3.5	1.2	3.5	1.3	3.2	1.5	3.3	1.4
Main government ministries and agencies										
MINAGRI—national	4.1	1.2	3.2	1.2	3.1	1.1	2.8	1.3	3.0	1.0
MINAGRI—local	3.2	1.1	2.5	1.2	2.5	0.9	2.3	1.1	2.6	0.9
MINRD—national	3.4	1.1	2.6	1.1	2.7	1.0	2.4	1.1	2.7	0.9
MINRD—local	2.9	1.4	2.3	1.0	2.7	1.0	2.3	1.2	2.4	1.0
Ministry of Land	3.8	1.2	3.0	1.3	3.1	1.3	2.8	1.4	3.0	1.2
MINREST	3.6	1.4	2.9	1.3	3.1	1.3	2.8	1.4	3.2	1.3
Ministry of Environment	4.1	1.3	3.4	1.4	3.5	1.3	3.1	1.5	3.3	1.3
Ministry of Finance	4.1	1.2	3.8	1.3	3.5	1.3	3.1	1.5	3.6	1.1
Ministry of Planning	4.2	0.9	3.6	1.2	3.6	1.1	3.3	1.4	3.7	1.0
Ministry of Infrastructure	3.8	1.0	3.3	1.1	3.4	1.1	2.9	1.2	3.1	1.3
Ministry of Women Affairs	3.6	1.5	3.3	1.8	3.4	1.4	3.0	1.5	3.3	1.4
Central Bank, Economic Research Dept.	4.5	1.1	4.2	1.0	4.0	1.2	3.8	1.1	4.4	1.0
Main platforms and associations										
CARG	3.4	1.1	2.6	0.9	3.0	1.1	2.8	0.9	3.3	1.2
FEC	3.9	1.2	3.2	1.1	3.6	1.1	3.4	1.1	3.6	1.1
COPEMECO	3.5	1.1	2.8	1.0	3.1	1.0	2.9	0.9	3.2	1.1
FOLECO	3.7	1.1	2.9	1.0	3.2	1.1	3.1	1.2	3.5	1.1
CNONGD	3.6	1.1	2.8	1.1	3.0	1.0	2.9	1.0	3.2	1.0
CRONGD	3.6	1.1	2.7	0.9	3.0	0.9	3.1	1.0	3.2	1.0

Source: Authors' compilation based on respondents. Notes: *Rating is based on a scale from 1 to 6 (where 1 = *no linkages* and 6 = *very strong linkages*). The figures are the mean scores of Net-Map interviewees and 18 respondents from different organizations in DRC, based on a semistructured questionnaire administered during the training on policy analysis on December 16, 2010.

Organizational capacity

Table 3 presents perceptions of stakeholders on the level of capacity of key organizations in the agricultural sector. Results suggest stronger human capacity in the universities especially UNIKIN, UCC, and University of Lubumbashi and in the Central Bank's Economic Research Department and the Ministry of Planning. Human capacity is weak at MINRD (national and local levels) and MINAGRI (local level). Of all the platforms and associations, CARG has the weakest human capacity while FEC has the highest score.

Unlike human capacity, the other areas of capacity (including facilities and infrastructure, organizational systems, M&E, and coordination and communication within organization) were rated to be weak or very weak by the respondents. Only UCC, Ministry of Finance, Ministry of Planning, Central Bank, and FEC have somewhat strong capacity in these areas. The weakest ones are the MINRD (national and local levels), MINAGRI (national and local levels), CARGs, Ministry of Land, MINREST, and Ministry of Women's Affairs.

Of particular interest are the universities and research organizations that provide evidence-based analysis and planning for the ministries. The university system faces numerous challenges including a disconnect between teaching, training and research on the one hand and policy problems and realities in the field on the other. Lack of investment and funding in university systems results in insufficient facilities, computer centers, and training materials and decaying human capital. Training of staff should be accompanied by upgrades in university infrastructure and equipment.

Human capacity

The current structure of the human resources in the agricultural sector is characterized by dichotomy of staff, with older staff at retiring age on one end and very junior staff with only bachelor's degrees or lower on the other. Qualifications and training of professional staff are generally low in MINAGRI, MINRD, and other key organizations in the agricultural sector. Hiring adequate staff and long-term training (master's and doctorate's degree programs) for junior staff will be required.

Analytical capacity

Human capacity in policy research and analysis is limited. Only about 13 PhDs—at UNIKIN, University of Yangambi, and UCC—are working on agricultural economics and rural development research for DRC's estimated population of 68 million. This figure is low compared with other African countries. For example, in Malawi, experts estimate around 50 PhDs working on agricultural economics and rural development research for an estimated 13 million population (IFPRI International Food Policy Research Institute 2012). There is a serious lack of policy analytical capacity at MINAGRI and MINRD. A diagnostic survey was administered to potential policy analysts in MINAGRI, MINRD, Ministry of Planning, Ministry of Land, INERA, Central Bank's Economic Research Department, and universities; and the 11 questionnaires that the team received back indicate limited knowledge in statistics and research methods, limited application of statistical techniques, and limited background in computer applications and software (Table 4). Those with a relative better perception of their proficiency are from the universities. The insufficient computer facilities contribute to this weak capacity of staff and their organizations. Substantive training as well as collaboration within existing

Table 4 Distribution of respondents based on proficiency in computer applications

Proficiency	No experience	Limited	Average	Good	Very good
Use of computer	0	3	3	3	2
MS Excel	0	5	0	5	1
Eviews	7	2	0	2	0
SPSS	5	5	0	1	0
Stata	9	2	0	0	0

Source: Author's compilation based on diagnostic evaluation of potential trainees on policy analysis.

capacity is needed for policy research. Training of staff should be accompanied by upgrades in infrastructure and equipment.

Provincial agricultural plans are the basis for a national plan. With decentralization, policy processes and decision-making are being conducted at the local level. It is important to build decentralized capacity for evidence-based policy analysis at the local level. Analytical skills for developing evidence based provincial plans and strategies are critically needed if the CARG approach to agricultural development is to succeed. While some capacity-strengthening activities exist at the national level (e.g., those conducted by FAO [FAO Food and Agriculture Organization of the United Nations 2008], the profound capacity gaps at the provincial level are yet to be addressed.

Policy dialogues and consultative processes

Several interviews with experts indicate serious problems in policy dialogue and consultation processes. First, consistently mentioned by all interviewees is the need for skills development on policy dialogue, communication, and negotiation. Second, key informants indicated lack of awareness and information on the status of reforms. One commonly mentioned example is the split between MINAGRI and MINRD, in which a number of respondents have some confusion on the division of functions. This hinders officials' and employees' understanding of their respective roles and responsibilities, limiting their incentive to perform. Some have concerns that the split will make coordination of agricultural and rural development policies on the ground more difficult. The reform coordination unit can include public awareness and reform communication strategies as an urgent activity. The extent of understanding and awareness by the public should also be one of the indicators of the success of the reform process that can be collected. The M&E system to be adopted by the reform coordination unit should be a result of an open and inclusive consultation. The communication platform and CARGs can be used as platforms for disseminating information about specifics and status of reform processes.

Monitoring and evaluation

M&E is one of the seriously neglected key processes in the agricultural sector. MINAGRI does not have a M&E system. The statistical office compiles data on agricultural production from territory levels. Each technical directorate will report this to the key officials of the ministry. The directorate of planning and studies at MINAGRI is tasked with M&E. Although there is one person in charge of M&E, there is no framework or plan in place. Other divisions of MINAGRI and other ministries confirmed that no work plans or performance-based evaluation system is in place.

Coordinated capacity development efforts

Key informants and a review of gray literature indicate numerous training and capacity-building efforts being conducted and implemented in DRC in various sectors including agriculture. While substantial donor resources are allocated to these trainings, still they are not institutionalized and have no sustainability mechanism. Agricultural training institutes and higher education institutes are often not involved in the training and capacity building being provided.

Organizational culture and incentives

Organizational culture in the agriculture organizations in DRC can be characterized by weak incentives to perform, fragile work environment, and lack of accountability as a result of the lack of capacity, incentive structures, and management systems described above. Table 5 provides a summary of the scores on perception of organizational culture. Respondents rated self-esteem and employee morale relatively high. Respondents also rated the quality of leaders and supervisors in their organization relatively high. Measures of transparency, fairness, political autonomy, coherence, openness, responsiveness, and flexibility were rated quite high on average, although answers vary widely among respondents. In contrast, measures of adequacy of resources, efficiency, freedom from corruption, job security, and mobility were rated low. Manifestations and sources of these disincentives among leaders and staff of key organizations include the following:

Weak political commitment

Agriculture is considered one of the "priority sectors" in government programs, in addition to education and health. In 2004, the government also promised at least 10 percent of public investments in agriculture. Despite this rhetoric, the budget allocated

Table 5 Perception of work environment, different organizations in DRC, 2010

Statement on work environment	Avg. score*	Std. dev.	Min.	Max.
You feel recognized by your boss and co-workers as a hard worker.	1.3	0.5	1.0	2.0
You are satisfied with your job.	1.6	0.6	1.0	3.0
The head of your organization is dynamic, inspirational, and respectful.	1.8	0.8	1.0	4.0
Your supervisor or boss knows enough about your daily activities to know if you are doing good or poor work.	1.9	0.5	1.0	3.0
Complaints from clients or partners are taken very seriously in this organization.	2.0	0.4	1.0	3.0
Your supervisor or boss consults you or asks your opinion regarding important changes.	2.0	0.8	1.0	4.0
There are good opportunities for promotion in your organization.	2.1	0.7	1.0	4.0
Your supervisor or boss gives you considerable opportunity for independence and freedom.	2.1	0.7	1.0	4.0
Performance evaluation in your organization is carried out in a fair way.	2.1	0.8	1.0	3.0
Male and female staff in your organization have equal opportunities in getting promoted.	2.1	0.8	1.0	4.0
There is hardly any political interference in our work.	2.1	1.1	1.0	4.0
The majority of people in your organization are well-qualified to do their job.	2.4	0.9	1.0	4.0
Your organization's staff are hired and promoted purely on the basis of merit.	2.4	0.8	1.0	4.0
In your organization, everyone has a clear understanding of their tasks and functions.	2.4	0.5	2.0	3.0
Clients or partners never complain about the performance of your organization.	2.5	0.7	1.0	4.0
Mobility to your operational area is easy.	2.6	1.0	1.0	4.0
Staff in your organization have to be worried about losing their jobs in the near future.	2.9	0.8	1.0	4.0
Corruption or misuse of funds is not a problem in your organization.	2.9	0.8	2.0	4.0
Your organization is effective given its budget.	3.0	1.0	1.0	4.0

Source: Authors' compilation based on respondents.
Notes: *Rating is based on a scale from 1 to 4 (where 1 = *strongly agree*, 2 = *agree*, 3 = *disagree*, 4 = *strongly disagree*). The figures are the mean scores of Net-Map interviewees and 18 respondents from different organizations in DRC, based on a semistructured questionnaire administered during the training on policy analysis on December 16, 2010.

to agriculture is well below the 10 percent recommended by the Maputo agreement. In 2007, the budget rate allocated to the agricultural sector was approximately 1.2 percent. It has decreased every year and has dropped to 0.7 percent in 2009 (MINAGRI reports cited by SADC Southern African Development Community 2009).

Lack of strategic direction and measurable targets

MINAGRI is still in the process of designing its overall agricultural strategy and plan, and these efforts have to be geared up within the new platform of CARGs. Unless such vision and concrete goals are produced, efforts to develop the sector will remain scattered and uncoordinated. The lack of human capacity at MINAGRI constrains effective agricultural planning and strategy formulation. The ministry has recently produced the DRC Agriculture Code and Agriculture Policy Note, which describe a general situation of the potential and constraints of the sector and provide an overall direction in terms of what the sector wants to achieve. But these documents do not contain any concrete goals or measurable targets. Furthermore, they do not consolidate efforts toward specific outputs and outcomes that can be validated, monitored, and evaluated over time.

Accountability measures

As already mentioned, MINAGRI has no M&E systems. At the CARG level, a document includes some indicators for CARG performance, but it pertains only to output indicators such as membership and attendance in meetings without mention of any outcome or impact indicators. The lack of information on tangible results to its members can hinder the sustainability and scaling up of good practices across CARGs. The reform coordination unit should review M&E systems of relevant organizations and make capacity-building for M&E systems a priority action.

Incentive constraints in CARGs

The formation of CARG shows both capacity and incentive issues. Areas of capacity gap range from negotiation and policy dialogue skills, organizational and management skills to competencies in marketing, processing, and value addition and M&E. Two meetings with CARG representatives by the study team and samples of meeting minutes reveal that key incentive problems faced by CARG formation include low attendance in meetings, weak connection with service providers, time and commitment of the executive secretary, sustainability of funds, and level of influence and power.

Individual incentives

Incentive for staff to stay in the public sector is limited. Pay is seriously low compared with the private sector and with the public sectors in other countries, which limits motivation and seriousness at work. High-level professionals earn less than US$20 per month in the provinces and a maximum of US$45 in Kinshasa, but the bulk of staff earn US$5 to US$15 a month (World Bank 2006). Field allowances generally are not paid.

Official statistics reveal that the nominal wage index in the private sector has increased by 23–78 percent from 2005 to 2007, while the public-sector wage index remained stagnant; and the real wage index in the private sector increased by 8–53 percent, while the real wage index in the public sector decreased by 11–17 percent during 2005–2007 (BCC Banque Centrale du Congo [Central Bank of Congo]). Wages generally are paid two to three months late; and some areas affected by armed conflict, such as North and South Kivu and Maniema, as of 2005 had not been paid for three years (World Bank 2006). MINAGRI can play a role in advocating for speedier civil service and salary reform.

Discussions and policy implications

A number of lessons emerge for strengthening the policy processes for agricultural development in DRC.

Investing inhuman capital and strengthening institutional linkages

The government must invest in hiring the minimum required number of staff and skill sets, minimum facility and infrastructure equipment, long-term training for public-sector staff, and designing and implementing evidence-based policies and strategies and monitor their progress and impacts. In the short term, a viable strategy to address the weak research–policy linkages would be the effective use of existing capacity. In the long term, with assistance from donors and international organizations, government should look into institutional arrangements for systematic linkages and sustainability of capacity-building efforts.

Strengthening monitoring and evaluation systems

Capacity building for designing and implementing M&E systems will be an important step toward greater focus on results, accountability, and performance-based planning and decision-making among relevant organizations. Public-sector organizations should set a vision and strategic direction to design their M&E frameworks and train staff in related skills such as data and statistical analysis and economic research to provide substantial, evidence-based strategic planning.

From technical assistance to organizational strengthening

Strengthening organizational capacity for public-sector organizations to perform more effectively will require technical assistance to design and manage financial and other systems, procedures, and resources. However, such technical assistance should focus on dual objectives of developing organizational procedures and the capacity for sustaining them.

Systemic assessment of capacity constraints and strategic investments

The design of capacity development investments need to adopt a dynamic and sustainable multi-stakeholder approach. This will need to include universities and colleges in the training programs and incorporate the training materials into their curriculum to ensure the continuity and institutionalization of training and learning efforts. The selection of individual trainees will have to be based on their ability to continue the training with other members of their organizations. A systematic assessment of the training landscape and the constraints and opportunities for the key actors and organizations is required before investing in capacity development.

Empowering farmers' organizations

There is a need to empower farmers' organizations to be strong voices and to advocate for the welfare of rural producers. Careful targeting of capacity-building efforts is critical to truly empower marginalized groups without promoting elite capture. The rural population will also need to improve their capacity to articulate their demands for rural services. Last, several incomplete elements in policy reforms such as clarity in fiscal allocations, definition of accountability measures, and civil-service reform need attention. The extent of public understanding and awareness is an indicator of the success of the reform process. Strategic communications are needed for informing the status of the reform process.

Conclusion

Achieving food security for all in developing countries requires appropriate institutional and policy reforms. However, the success of policy and institutional reforms is intricately linked to the capacity of the organizations and actors involved in these processes. Yet, applied studies on understanding the organizational and capacity constraints and challenges are few and far between. This is partly due to the single-disciplinary treatments of policy processes that remain the domain of political economists - mostly in developed countries - and the study of institutional, organizational, and capacity challenges by public administration and social scientists. Yet, understanding and solving institutional and policy systems challenges require multidisciplinary approach to the study of institutions that are context and time specific. Limited applications to developing country context is also due to lack of analytical and data collection methods to document the policy processes, institutional change, and organizational and individual capacities.

This paper brought the policy process and organization development disciplines to develop a conceptual framework to trace the levels, linkages, and influence of various organizations and actors in the policy process. It applied this framework to study the institutional and policy reforms that are currently underway in DRC, a country still slowly emerging from war and conflicts. Using an innovative methodology to map the policy process with a view to analyze the organizational and capacity challenges it identified various capacity and incentive bottlenecks in the agricultural sector.

Results indicated that moving the policy and institutional reform agenda forward will require higher level of political commitment, increased investment support, systematic capacity development at the organizational and individual levels, functioning monitoring and evaluation system, and improved research-policy linkages. A key lesson emerging from the case study is that while local ownership of the policy and institutional reform process has to be nurtured, allowing local policy makers to solve their problems in their own context is critical. Identifying opportunities for guiding the reform processes and providing strategic support to the existing local organizations by the external funders should begin from what is available within the country and by building upon them. The role of evidence-based analysis by developing the capacity of the local researchers in furthering institutional and policy reforms can hardly be overemphasized.

Endnotes

[a]Policy process is defined in this paper as the process of public policymaking where problems are conceptualized, discussed and debated, alternative solutions

developed, policy choices are made, implemented, monitored, evaluated and revised (Sabatier 2007).

[b]Capacity considered in this paper are human or individual (skills, knowledge, education, and training of individuals); organizational (physical and financial resources, and human resource and management systems in the organization); and institutional (work environment and culture, staff morale, motivations, and accountability within organization and across organizations).

[c]Capacity gaps are defined in this paper as discrepancy between the expected or required capacity to achieve a set goal and actual or observed capacity.

[d]MINRD was first created in the 1980s, but it was absorbed twice by MINAGRI and then re-created as an independent ministry.

[e]Based on Decree Law 081 (July 2, 1998) on territorial and administrative organization of the Democratic Republic of Congo. This figure includes three districts in Kinshasa (with no chief of district). Most of the districts are to become provinces (except the newly created district of Plateaux), according to Article 2 of the Congolese Third Republic Constitution.

Abbreviations

BCC: Banque Centrale du Congo [Central Bank of Congo]; BTC: Belgian Technical Cooperation; CAADP: Comprehensive Africa Agricultural Development Programme; CARGs: Agricultural and Rural Management Councils; COPEMECO: The Confederation of Small and Medium Enterprises of the Congo; DRC: Democratic Republic of Congo; FAO: Food and Agriculture Organization; FEC: Federation of Private Enterprises in Congo; FOLECO: Federation of Liac and Economics; GDP: Gross Domestic Product; IFPRI: International Food Policy Research Institute; INERA: National Institute for Agricultural Research and Studies (Institut National des Etudes et de la Recherche Agricole); MINAGRI: Ministry of Agriculture, Fisheries, and Livestock; MINRD: Ministry of Rural Development; MINREST: Ministry of Scientific and Technical Research; NGOs: Non-Governmental Organizations; NSSP: Nigeria Strategy Support Program; ODI: Overseas Development Institute; SADC: Southern African Development Community; UCC: Catholic University of Congo; UNIKIN: University of Kinshasa; USAID: United States Agency for International Development.

Competing interests

The authors declare that they have no competing interests.

Authors' contributions

CR, SCB and JU conducted the interviews and fieldwork together. CR conceptualized the framework and methods, analyzed the initial results and drafted the first version. SCB contributed on the literature review, framework and discussion sections. JU provided information and sections on the DRC context, made arrangements for the interviews to various officials and key informants in DRC, and added on the discussion and policy implications. All authors read and approved the final manuscript.

Acknowledgements

The authors would like to thank Cornelie Sifa Nduire and Joel K. Siku. The authors are very grateful for the invaluable guidance and comments from Regina Birner. Thanks go also to the DRC government officials and private individuals who shared data, information, and experiences, and the many representatives from various organizations whose active and dynamic participation in the interviews and focus groups gave rich resources and insights to this paper.

Author details

[1]Development Strategy and Governance Division, International Food Policy Research Institute (IFPRI), 2033 K Street, NW, Washington DC 20006, USA. [2]Partnership, Impact and Capacity Strengthening Unit, International Food Policy Research Institute (IFPRI), 2033 K Street, NW, Washington DC 20006, USA. [3]West and Central Africa Office, International Food Policy Research Institute (IFPRI), 2033 K Street, NW, Washington DC 20006, USA.

References

Aberman N, Schiffer E, Johnson M, Oboh V (2010) Mapping the Policy Process in Nigeria: Examining Linkages Between Research and Policy. International Food Policy Research Institute (IFPRI) Discussion Paper 01000. IFPRI, Washington, DC

Adebayo K, Babu SC, Rhoe V (2009) Institutional Capacity for Designing and Implementing Agricultural and Rural Development Policies and Strategies in Nigeria. Nigeria Strategy Support Program (NSSP) Background Paper No. NSSP 008. IFPRI, Washington, DC

Ayuk ET, Marouani MA (2007) The Policy Paradox in Africa: Strengthening Links Between Economic Research and Policymaking. International Development Research Centre/Africa World Press, Ontario, Canada

Babu SC (2013) Policy Process and Food Price Crisis: A Framework for Analysis and Lessons from Country Studies. Working Paper 2013/070. World Institute for Development Economics Research, Helsinki

Babu SC, Mensah R, Kolavalli S (2007) Does Training Strengthen Capacity? Lessons from Capacity Development in Ghana, Ministry of Food and Agriculture. Ghana Strategy Support Program (GSSP) Background Paper No. GSSP 0009. IFPRI, Accra, Ghana

BCC (Bangue Centrale du Congo [Central Bank of Congo]) (2004, 2005, 2006, 2007) Annual Report. BCC, Kinshasa

Booth D (2011) Aid, institutions and governance: what have we learned? Dev Policy Rev 29(S1):S5–S26

Cash DW, Clark W, Alcock F, Dickson NM, Eckley N, Guston DH, Jager J, Mitchell RB (2003) Knowledge systems for sustainable development. Proc Natl Acad Sci 100(14):8086–8091

Court J, Young J (2003) Bridging Research and Policy: Insights from 50 Case Studies. Overseas Development Institute (ODI) Working Paper 213. ODI, London

FAO (Food and Agriculture Organization of the United Nations) (2008) Evaluation of FAO Cooperation in Democratic Republic of Congo. Executive Summary, Kinshasa, FAO

Grindle MS (2007) Good enough governance revisited. Dev Policy Rev 25:533–574

Guston DH (2001) Boundary organizations in environmental policy and science: An introduction. Sci Technol Hum Values 26(4):399–408

IFPRI (International Food Policy Research Institute) (2012) Food Policy Capacity Indicators. Draft Report, IFPRI, Washington, DC

Killick T (2010) Development Economics in Action: A Study of Economic Policies in Ghana, 2nd edn. Routledge, London

Marshall J, McLean A (1988) Reflection in Action: Exploring Organizational Culture. In: Reason P (ed) Human Inquiry in Action. Sage Publications, London

Omamo S (2004) Bridging Research, Policy, and Practice in African Agriculture. Development Strategy and Governance Division Discussion Paper 10. IFPRI, Washington, DC

Resnick D, Birner R (2010) Agricultural strategy development in West Africa: The false promise of participation? Dev Policy Rev 28(1):97–115

Rivera W, Alex G (2004) Volume 2. Privatization of Extension Systems: Case Studies of International Initiatives, Agriculture and Rural Development. Discussion Paper 9. World Bank, Washington, DC

Ryan J (1999) Assessing the Impact of Rice Policy Changes in Vietnam and the Contribution of Policy Research. Impact Assessment Discussion Paper 8. IFPRI, Washington, DC

Sabatier PA (2007) Theories of the Policy Process, 2nd edn. Westview, Cambridge, MA

SADC (Southern African Development Community) (2009) SADC Regional Agricultural Policy (RAP)—National Review Report. Democratic Republic of Congo, Draft

Schiffer E (2007) The Power Mapping Tool: A Method for the Empirical Research of Power Relations. Discussion Paper 00703. IFPRI, Washington, DC

Stone D, Maxwell S, Keating M (2001) Bridging Research and Policy. Working Paper. Warwick University, Coventry, UK

Tollens E (2004) Les défis: Sécurité alimentaire et cultures de rente pour l'exportation - principales orientations et avantages comparatifs de l'agriculture en R.D. Congo, Table Ronde sur l'Agriculture en RDC, "Vers une stratégie de développement agricoles, base solide du décollage économique, Kinshasa", 19-20 Mars, 2004, Cercle de l'Alliance Belgo-Congolaise, 2004. Kinshasa: Cercle de l'Allinace Belgo-Congolaise, p 32

World Bank (1999) Capacity Building in the Agricultural Sector in Africa. World Bank Operations Evaluation Department, Spring 1999, No. 180, Washington, DC

World Bank (2006) Democratic Republic of the Congo: Agricultural Sector Review. Main Report, April 15. Report No. 30215-ZR. Washington, DC: World Bank.

World Bank (2010) Designing and Implementing Agricultural Innovation Funds: Lessons from Competitive Research and Matching Grant Projects. Washington, DC

Young J (2005) Research, policy, and practice: Why developing countries are different. J Int Dev 17:727–734

4

Performance and profit sensitivity to risk: a practical evaluation of the agro-industrial projects developed by Israeli companies for the CIS and Eastern European countries

Gregory Yom Din[1,2]

Correspondence:
gregoryyd@gmail.com
[1]Department of Management and
Economics, the Open University of
Israel, Raanana, Israel
[2]Faculty of Exact Sciences, Tel-Aviv
University, Tel-Aviv, Israel

Abstract

International companies take part in many tenders for agro-industrial projects in the Commonwealth of Independent States and Eastern European countries. The market for these projects is analyzed and found to be favorable for companies and developers. Major projects developed in recent years are presented and evaluated in terms of financial performance. Additionally, a method of project evaluation by profit sensitivity to risk criterion is proposed. In this method, the approximate formula for profit sensitivity to risk (when basic production and market assumptions change simultaneously) is derived using a cost-volume-profit model. This method allows minimal calculations to explain profit sensitivity and elasticity within the usual indicators of business planning: operational profitability and degree of operating leverage. The consistency of project ranking is examined using Cronbach's alpha and correlation coefficients. The ranks obtained by various performance criteria are found to be consistent with each other, but not with those obtained by profit sensitivity to risk. In terms of elasticity, project profitability is a much stronger influence than the degree of operating leverage on profit sensitivity to risk.

Background

Many Commonwealth of Independent States (CIS) and Eastern European (EE) countries have enjoyed a period of rapid economic growth since the beginning of 2000's, except for the 2008–2010 period of financial crisis. During 2000–10, investment in fixed assets in CIS countries increased by 150 percent in terms of real prices, in Romania by 122% and in Bulgaria by 66%. In comparison, fixed asset investment in EU-27 countries averaged an 18% increase over the same period. In Russia, this investment reached US $290 billion, with the agro-industrial sector accounting for 4%. Over 2008–11, the annual investment in Russian agriculture was about US $12 billion, in Ukraine $2 billion, in Belarus $2 billion, in Kazakhstan $620 million and in Romania $940 million (in 2009) (RosStat 2012, UkrStat 2012, BelStat 2012, KazStat 2012, Romania Stat 2012, EuroStat 2012b).

Many international companies compete in this market and submit projects for potential implementation. In Russian agriculture, international and joint domestic-international companies accounted for 6.5% of investments over 2009–10 (RosStat 2012). Every year, Israeli companies are involved in dozens of tenders for agro-industrial projects in various CIS and EE countries for the industries of poultry, dairy and pork production, fish farming, vegetable greenhouses, orchards, fruit packing houses and oilseed products. In order to win tenders, companies invest heavily in the development of economic models and business plans in accordance with accepted criteria, government guidelines and banking standards.

The objectives of this study are three-fold:

a) To describe the market situation for agro-industrial projects in the studied countries from the viewpoint of project developers.
b) To propose a method of project evaluation using profit sensitivity to risk analysis, in conjunction with evaluation by financial performance criteria.
c) To evaluate agro-industrial projects developed in Israel according to these criteria, and to analyze ranking consistency by different criteria.

A number of studies evaluating agricultural profitability in CIS and EE countries have been published in recent years. In the 1990's, most Russian companies considered the agricultural sector to be unprofitable and interest from foreign investors was limited. During this period, the Russian government sharply curtailed investment in the agricultural sector and large farm enterprises faced severe financial difficulties (Visser et al. 2012). Voigt and Hockmann (2008) found that from 1993–2003, there was little evidence documenting sustainable growth in Russian agriculture, and no significant transition progress was revealed. Agricultural production was rather industrialized but operated with decreasing returns to scale.

After 1991, agricultural systems in CIS and EE countries underwent major institutional changes. In the largest countries – Russia, Ukraine, and Kazakhstan – 23 million hectares of arable land was excluded from production by the mid-2000's (Lioubimtseva and Henebry 2012). These losses in production capacity were not accompanied by an adequate improvement in management and technology.

Liefert and Liefert (2012) addressed issues of agricultural productivity in Russia, Ukraine and Kazakhstan in this period of economic transition, and studied the effects of modern technology and management techniques on improving agricultural productivity and maximizing profits. They found that the majority of large agricultural enterprises remained technologically backward and chronically unprofitable, despite their industrial nature and vertical integration. Nevertheless, these large agricultural holdings played a large part in production for the large CIS countries in the end of 2000's. They were responsible for approximately 15% of total agricultural production in Russia, accounted for 66% of Kazakh grain marketed domestically and abroad, and cultivated 10% of the total farmland in Ukraine (Wandel et al. 2011).

Welfare of small farmers in many CIS and EE countries depended mostly on the market price for their output in environments where agricultural policy provided minimal support. These farmers could suffer from price heterogeneity even after controlling for product heterogeneity, as it is shown by Sauer et al. (2012) in analyzing the dairy sector of Armenia, Moldova and Ukraine. Voicilas (2011) cites high risk, weak profitability and institutional failures (slow pace of privatization, negligible reforms and high taxation) as the main reasons for the minimal place of Romanian agriculture in foreign direct investment. In 2008, only 1% of foreign investment was in agriculture, although that sector employed 30% of the total workers in the country.

In the literature on project evaluation, the choice of modeling tools and criteria has an important place. Production scheduling, sales planning, and cash budgeting are integrated in spreadsheet models through linking procedures, allowing for the development of an effective business plan (Chien and Cunningham 2000). For industrial agricultural projects, the dynamics of the growing process in crop industries and of herd/flock movement can be entered into a bio-economic model spreadsheet, in which the output of the biological and production processing modules serves as an input to an economic (cash flow) module.

Bio-economic framework developed as an Excel-based representation allowed for an evaluation of profit and resource management at a project/farm business level (Zhang and Wilhelm 2011, Kuehne et al. 2012). The problem of criteria in project evaluation was studied by Parfenova (2009) who discussed socio-economic criteria for industrial agricultural projects aimed at regional economic growth and improving general living standards. Mansurov (2011) proposed different criteria for competitive evaluation of industrial agricultural companies by decision-makers such as project initiators, business owners and government organizations. The latter are particularly interested in increased tax revenue and in tools for solving social problems. The methodological aspects of project evaluation, particularly the determination of the discount rate for cash flows, was analyzed by Bevzelyuk (2008).

Many of these studies incorporate the issue of risk in project evaluation. Strashko (2010) analyzed issues of business planning in relation to agro-industrial projects, specifically the need for risk, technical and financial analysis during the project evaluation. Hockmann et al. (2011) studied the influence of risk, expressed as variation in production and prices, on agricultural development and production growth in one of the regions of central Russia. The dual nature of risk in agriculture that follows from output and price uncertainty was analyzed by Ben-Zion et al. (2005), in the context of analyzing the difficulty in hedging agricultural production. On the other side, large-scale agro-industrial projects allow for export diversification in CIS countries, by replacing raw materials (mainly grain) with finished products. This enables the reduction of risk introduced by price volatility in the world market (Shepotylo 2012; Ksenofontov et al. 2012).

This brief review highlights the importance and specific details used for evaluating agro-industrial sector investment projects in the studied countries – the role of social factors, participation of multiple decision-makers, and exposure to multiple risk factors. However, practical examination of various evaluation methods for agro-industrial projects is not thoroughly presented in current economic literature. In particular, it is of great interest to compare result consistency of evaluations performed with different criteria and methods. Such examination is important, both for project developers and for customers interested in seeing multi-criteria investment project evaluation. In Russia, for example, with its technological and economic stratification in the agricultural sector, there are no mechanisms for harmonizing the interests of key decision-makers in regards to the evaluation of projected agribusiness projects and their solvency (Zaharov 2006; Kalugina 2011; Vasina 2012).

The results of this study are based on data collected by the author in the course of more than fifteen years of work for Israeli companies on economic modeling and business planning for agro-industrial projects.

The market situation for agro-industrial projects

Food demand and production

In the last decade, agricultural production in major CIS countries has increased significantly. From 2007–10, as compared to the period of 2003–06, the Food Production Index grew in these countries at a rate of 2.4-5.3% per year. In other countries in the region there was a mixed trend. In the Baltics, Estonia saw a decrease in production (–1.8%) while significant increases were shown in Latvia (5.6%) and Lithuania (3.2%). In many EE countries, the average annual growth rate was negligible, such as in Bulgaria (1.2%), or even negative, as for example in Romania (–1.6%). In Israel, the average annual growth rate was negative at –0.5% (Figure 1) (Commodity Markets 2012).

Many studies suggest that by 2050, the worldwide demand for food will have increased by some 70-100% (Wise and Murphy 2012). In line with this outlook, and according to the more attractive features of agro-industrial export as opposed to traditional agricultural products (Torok and Jambor 2012), the amount of agro-industrial projects will undoubtedly increase in the countries studied. They have the necessary production potential for this to take place. In the three largest CIS countries – Russia, Ukraine and Kazakhstan – the potential for wheat production, evaluated on the basis of current yield totals, is about 99 million tons per year, which exceeds the current total production of wheat in the United States and Canada combined (based on data from Petrick et al. 2012). Agricultural exports to the European Union increased from 2003–07: from Russia by 24%; from Ukraine by 77%; from Kazakhstan by 102%; from Belarus by

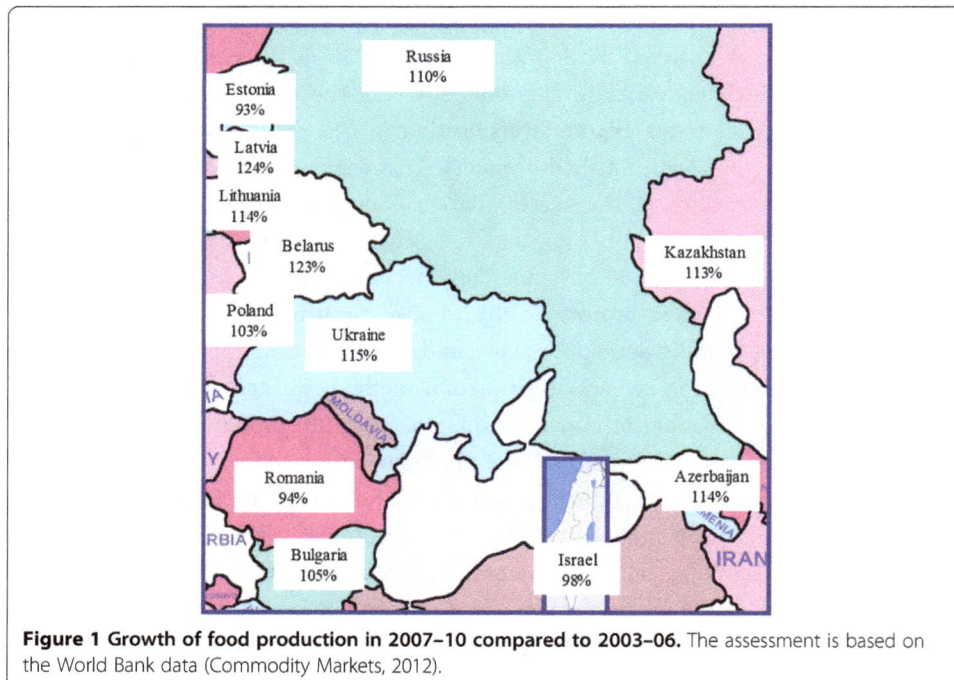

Figure 1 Growth of food production in 2007–10 compared to 2003–06. The assessment is based on the World Bank data (Commodity Markets, 2012).

10%, from Romania by 51%, and from Bulgaria by 40%. The total annual value of agricultural exports from CIS countries to the EU reached US $7.8 billion in 2012, with Russia accounting for almost half (based on data from Bojnec and Fertő 2012).

Price growth

In recent years, the global market for agricultural products has been characterized by a sharp rise in producer prices. The index of world prices for agricultural products grew during 2000–11 by 7.1% per year (FAOSTAT 2012). Twice during this period, in 2008 and 2010, the index increased particularly sharply: by 19% and 16%, respectively. This led to media reports of a world food crisis. In individual countries, prices also have risen sharply. For example from 2005–09 in Romania, the price index of basic agricultural products grew by 38% (Romania Stat 2012).

In 2011 the growth in world prices slowed, and the index rose by only 7%. However, for many categories of food products that are relevant for CIS and EE countries, the price index has continued to grow rapidly. In 2011, the world price index for wheat in terms of producer prices increased by 33%, for soybean oil by 29%, and for beef by 21%. Part of the growth in producer prices can be explained by increased prices for many resources important to agriculture – in 2011, energy prices rose by 19% and prices of chemical fertilizers by 43% (FAOSTAT 2012).

Availability and affordability of food

The industrial agriculture market is influenced by governmental commitment to food security. Current understanding of this problem was formulated by the 1996 Rome Declaration on World Food Security. In Russia, this term was defined in 2010 in the "Doctrine of Food Security," in Ukraine by a 2011 law, in Belarus by a 2004 government decision, and in Romania by the document "Strategies for Food Security," presented to the European Commission in 2001. The following characteristics of food security are defined in these documents: food quality, stable and sufficient production, and affordability for all residents.

The share spent on food from a household budget, as based on Engel's law, is one of the most important aspects in assessing standard of living. A decrease in food expense relative to total budget particularly indicates improved food affordability. In recent years, this index in Russia remained close to 30%, similar to 1985 levels. In Ukraine in 2008 (a very unfavorable year economically) the index was even higher: 51%. In 2010, Belarus and Kazakhstan saw household food budgets of 39% and 41%, respectively. In 2005, the EE countries of Romania and Bulgaria had rates of 44% and 32%, respectively, and the EU-27 had an average rate of 17% (RosStat, UkrStat, BelStat, KazStat, Romania Stat 2012, EuroStat 2012a).

These estimations are in line with findings of other researchers. Otiman (2012) concluded that in Romania, the share of food expenses was between 40-45%, a value twice as high as the EU-25 average and almost 2.8–3 times higher than the EU-15 average.

These data indicate that the relative cost of food in a household budget in the countries under consideration is 75-160% higher than the cost in EU countries, and thus affordability of food is much lower.

The benefits and risks of agro-industrial projects

Modern technologies that are used in agro-industrial projects often belong to the field of agricultural biotechnology. This can be discerned in the areas of production, processing, storage, organization, and management. Long-term storage of fruit in controlled environment, sex changes of fish in aquaculture, protein production in soybean processing plants and extraction of industrial oils from biomass are all examples of this.

Products from agro-industrial projects are notable for their high, uniform quality and for their suitability for delivery to distant markets. Meat products from the poultry industry in Belarus is one example of this. Chicken pieces shipped in user-friendly, high-quality packing have a third more value than a whole bird, and are delivered to major cities in Russia where prices are much higher as compared to Belarus (Table 1).

In new companies established for project implementation, food production is stable even in years of adverse weather conditions. For example, Russia in 2010 saw a 25% decrease in general crop production because of low yields in many regions. At the same time, meat production increased by 6%, because the industry is largely concentrated in industrial farms. In part, this increase can be explained due to feed shortage (based on data from RosStat 2012).

Many potential risks of industrial agricultural projects are derived from their high cost of implementation and profit volatility. Profitability is affected by inertia in international agricultural markets that are affected by trends of the previous year. Another feature that affects profitability is long production cycles and the initial period needed for raising a main herd, parent flock or fruit orchard. In the poultry sector, the period from the beginning of investment until the first revenues is at least 14 months, and in fruit orchards the period is 4 years, with another 2 years to reach full market yield. A similar long period is required to achieve full production from a new dairy farm.

Agro-industrial projects require initial investment and working capital relatively high as compared to traditional agricultural projects. Small-scale horticultural enterprises require US $5-7 million, while large integrated poultry farms require investment of up to US $50–90 million or more (Table 2). In CIS and EE countries, price volatility in the agricultural sector is much higher than yield volatility. From 2000–11 in Russia, the coefficient of variation, which measures the standard deviation as a part of the sample

Table 1 Wholesale prices of poultry meat products

Product		Prices*, US$/кг, and differences	
		Turkeys	Chicken
Belarus			
Whole bird	B1	4.67	2.06
Fillet	B2	6.51	2.81
Price difference		139%	136%
Russia			
Whole bird	R1	5.75	2.67
Fillet	R2	7.78	4.48
Price difference		135%	168%
Difference between	R1/B1	123%	130%
Countries	R2/B2	120%	159%

*The assessment is based on price proposals published from November 2011-January 2012.

Table 2 General project characteristics

№ and year	Project	Country, region	Products	Investment 2012 US$ million
1 2004	Eggs	Russia, Central FD*	Eggs	44
2 2006	Turkeys	Russia, Privolzhsky FD	Meat products	86
3 2009	Fruit	Azerbaijan	Apple from storage	5.3
4 2009	Vegetable greenhouses	Azerbaijan	Tomatoes	7.0
5 2009	Broilers	Russia, Privolzhsky FD	Chickens	80
6 2009	Broilers	Kazakhstan	Chickens	53
7 2010	Eggs	Romania	Eggs	27
8 2010	Orchard+ oil extraction	South-Eastern Europe	Technical oil	1.5
9 2011	Pig farm	Belarus	Meat, meat products	31
10 2011	Turkeys	Russia, Central FD	Meat, meat products	110
11 2011	Turkeys	Belarus	Meat, meat products	87
12 2011	Soy	Russia, Far East FD	Oil, soy proteins	74
13 2012	Eggs	Belarus	Eggs, melange	28
14 2012	Milk farm	Russia, North-West FD	Milk, cattle	42

* FD - Federal district.

mean, was 16% for grain crops yield and 40% for the price of these crops (based on data from RosStat 2012).

Because of the mentioned factors, using profit sensitivity as a criterion of agro-industrial project analysis is justified. For the purposes of this article, risk is defined as simultaneous changes in several basic production and market factors that affect profitability and project investment.

Institutional context in project choice and evaluation

The institutional environment of the agricultural industry in the countries under consideration has changed decisively over the last decade. Below the most important changes in Russia are outlined; similar changes have taken place in other studied countries.

In 1992, Russia introduced a set of reforms aimed at building a market economy. The reformers predicted that the necessary institutions would come into being after private property was created (Goldman 2003). However, the results stemming from changes in land management were unexpected for the reformers. The absence of an adequate institutional environment led to a falloff in volume of agricultural production so drastic that the food security of the country came into question (Kalugina 2011).

After 2000, agricultural regulation and budgetary support was decentralized, and the center of gravity for agrarian policy was shifted from the central government to the regions (Saubanov 2010). The emergence of bank-issued credit served as an important engine for economic growth, and it spread from the Moscow district to other Russian regions (Berkowitz and DeJong 2011). These institutional changes led to a major transformation in the agro-industrial sector: private land ownership on a large scale, increased ability for producers to respond to market conditions and demand, creation of functioning wholesale markets for raw materials and agricultural products, an increased number of large integrated farms, and additional entrepreneurship opportunities. Investment environment in the agro-industrial sector were improved due to state support of investments in transportation infrastructure, cold storage, distribution, access to foreign markets for equipment and genetic material, interest rate subsidies, tax benefits, and customs preferences.

In Russia, further expansion of the state support in the development of agriculture and regulation of agricultural commodities markets is planned for 2013–2020 (The State Program for Development of Agriculture and Regulation of Agricultural Commodities Markets in 2013–2020 2012). The budget of the program accounts for about 16 US$ billion divided by industries of crop and livestock sectors, land improvement, innovation, support in small farms, and social development of rural areas. The wide public support reduces the overall level of risk for the new projects.

These changes contribute to the development of the industrial agricultural sector. In 2010–11 in Russia, 70% of agricultural land that was in use was under private ownership (Wegren 2012). Every month, more than 500 new companies register in the agricultural sector (RosStat 2012, including hunting and forestry enterprises). In 2009, total investment credit given by the Russian foreign trade and investment bank stood at US $7.7 billion, with 25% going to agro-industrial projects (Isakov 2011).

The following actors are central in evaluating agro-industrial projects, and in the decision to grant them subsidies and benefits:

a) The *region*, including the governor and the relevant departments of the regional government;
b) The *bank*, which participates in project financing and has experts who influence decisions;
c) The *initiator* of the project, whose interests usually coincide with those of the future owners of the new company, which will be established during project implementation.

These actors examine the project business plan and other relevant documents and perform due diligence by analyzing economic performance and investment risks. The *region* often seeks the opinion of academic institutions, the *bank* has internal experts to examine a business plan, and the *initiator* tends to seek the services of a reputable consulting firm.

Criteria for the *region* when deciding whether to support a project are often based on projected profitability for the new company – the budget effect of additional taxes for the *region* – and on the opportunity to obtain additional financing from the central government to invest in infrastructure supporting a profitable project. Social concerns are also important for the *region*, such as food security issues, employment opportunities, and new positions for workers with professional education. Besides providing administrative support, the *region* can decide on a substantial reduction in asset tax (ranging from 2.2% to 0.5%). In some countries, the *region* can subsidize interest rates.

The bank uses the same performance criteria as the *region*. In addition, the *bank* is particularly attentive to analysis of profit sensitivity from project assumptions. This gives information about loan risk. Through the *bank*, a state interest rate subsidy can be implemented – in some cases, 75% and more of the interest can be returned to the new company. Based on the project's financial forecast, the *bank* can grant a 2–3 years grace period before for principal loan repayment begins.

The initiator pays relatively more attention to criteria based on discounted cash flow, especially the IRR. This performance measure allows for comparison between various project proposals. An additional concern is the ability to profit from vertical integration if a new project can use agricultural raw materials produced from enterprises already owned by the *initiator*. The *initiator* is responsible for mobilizing equity – usually, 20-25% of investment cost. Examples of how much the *initiator* can be motivated in his decisions on choosing an agricultural project and in the production of specific agricultural products in the last decade are as follows: Reduced single taxes for agricultural producers to 6% in Russia and 1% in Belarus. Subsidy for specific products can play a part: in Russia, a state subsidy for the high quality milk can reach 25% of the milk farm gate price, and in Romania, an egg farm is granted 2.5 euro per layer per annum.

Methods

Data

The project evaluation data were collected and evaluated as follows:

- Technological data related to the production processes were provided by the Israeli companies that developed the projects, and by the suppliers of genetic material, equipment, and construction material. Local specialists were consulted regarding the possible impact of agro-climatic conditions on the projected technological data, e.g. soybean humidity, seasonal variation in egg lay rate, or feed nutritional value.
- Costs of labor, raw materials, energy, and services were collected from local statistics reports and annual reports of corporate customers. Labor costs, direct and indirect, included wages, state benefit fund payments and insurance. For each category of employees, their wages were projected 20-50% higher than the regional average. Feed costs were estimated using regional prices of fodder and feed additives and recommended optimal rations for different animal groups.
- Producer prices of products were estimated from data acquired from central statistical services, reports of various ministries of agriculture, and letters of intent from trade organizations and processing plants. Conservative price estimates (1–2 years average) were used, differentiated by region, season, and production quality.
- Construction costs were estimated based on tenders of local contractors and on data collected during work visits to project sites.

Costs and prices were estimated in euros or in US dollars. Neither was extrapolated, but their linear 10 year trends were used in sensitivity analysis, for estimating possible changes in model assumptions.

Of the 14 projects analyzed in this study, six projects were developed for various regions of Russia, six for other CIS countries, and two for EE countries. All the projects relate to agricultural production and to storage/processing/packaging.

Processing agricultural raw materials included analysis of the following stages: production of pig and poultry meat products, production of feed mixes in livestock farms, production at soybean processing plants, oil extraction from plant material, long-term storage of fruit in controlled atmosphere, and egg product production. The average cost of investment in 3 projects based on crops (fruit, vegetable greenhouses, and plantation and oil extraction) was 4.6 million US $. For a soy processing project, which requires the construction of an oil extraction plant, other production units, and storage bins for raw material, 74 million US $was invested, and for 10 projects based on livestock production the average investment was 59 million US $ (Table 2).

Technological characteristics of the projects are shown in Table 3, and compared to the countries' average or typical values. These comparisons change in various industries; the greatest difference is in the vegetable greenhouse industry, and close to zero differences are shown in the soy industry, broilers, and the egg industry in Russia and Belarus. In contrast, in the same egg industry in Romania the difference is essential. The table footnotes include a list of data sources used for comparing the projects' characteristics to the countries' average or typical values. In Table 3, most projects show high vertical integration.

Using spreadsheet models for business planning

For evaluation purposes, a spreadsheet business plan model was developed for each project. Production was detailed by biological/technological phases. For an integrated poultry meat farm, for example, these phases could include rearing the parent flock, breeding, hatchery, brooding and fattening commercial chicks, slaughtering and meat processing. Feed expenses for livestock farms were described in a separate module. Weekly step models were used for the poultry farms, and monthly step models for other livestock farms and crop industries. Cash flow was forecast based on output from production and sales modules. These flows included modules of income and expenses, along with production, investment, and financing cash flows. The planning horizon was assumed as the sum of the initial development period plus additional 5 years of production activity. For plantations, a longer planning horizon was used.

The models allowed for the calculation of all economic indices used in project evaluation:

- Variable and fixed costs per unit of product;
- Operating profit;
- Degree of operating leverage; and
- The variety of economic performance measures as they relate to nominal and discounted cash flows, current earnings and investment.

Using equations derived from herd and flock movement (in terms of weeks and months) and from plantation/greenhouse productivity (in terms of months and years) enabled sensitivity analysis of project profitability, based on spreadsheet formulae. For various industries, the following parameters of sensitivity were used: production prices in farm gate value; production volume; total production costs, production costs of feed

Table 3 Technological characteristics and vertical integration of the projects

№	Project, country, Federal District	Major technological characteristic				Vertical integration
		Value	Unit	Compared to country's value	Data source for the comparison	
1	Eggs, Russia, Central	308	Eggs/layer/year (lay rate)	101%	A	Parents flock, hatchery, feed mill
2	Turkeys, Russia, Privolzhsky	2.62	Feed Consumption Ratio	61%	B	Parents flock, hatchery, feed mill
3	Fruit, Azerbaijan	38.3	Apple ton/ha	43.2%	C	Pack-house, fruit long storage in controlled atmosphere
4	Vegetable greenhouses, Azerbaijan	485	Tomato ton/ha/year	513%	D, E	Growing seedlings, packaging production
5	Broilers, Russia, Privolzhsky	2.0	Feed Consumption Ratio	98%	F	Parents flock, hatchery, feed mill
6	Broilers, Kazakhstan	2.0	Feed Consumption Ratio	98%	F	Parents flock, hatchery, feed mill
7	Eggs, Romania	308	Lay rate	224%	G	Parents flock, hatchery, feed mill, manure pelleting plant
8	Orchard+extraction	537	Kg of tea tree oil/hectare	363%	H	Plantation, oil extraction plant
9	Pig farm, Belarus	3.23	Feed Consumption Ratio	77%	I	Main herd, feed mill, meat products plant
10	Turkeys, Russia, Central	2.39	Feed Consumption Ratio	56%	B	Hatchery, feed mill, meat products plant
11	Turkeys, Belarus	2.62	Feed Consumption Ratio	61%	B	Parents flock, hatchery, feed mill, meat products plant
12	Soy, Russia, Far East	16%	Oil yield	102%	J	Seed storage, soy processing, bottling line, feed mill
13	Eggs, Belarus	308	Lay rate	102%	K	Parents flock, hatchery, feed mill, egg melange plant
14	Milk farm, Russia, North-West	11	Ton of milk per cow/year	262%	A	Main herd, feed center, biogas plant

A) Russia's average in 2008-2010, for agricultural enterprises (RosStat 2012).

B) Turkey of Stavropol region (2012). Recommendations of this breeding center, the largest in Russia, are taken as a base for comparison for projects both in Russia and Belarus.

C) FAOSTAT (2012), yield in Azerbaijan, 2008-2010: 8.87 ton/ha.

D) FAOSTAT (2012), yield in Azerbaijan, 2008-2010: 17.1 ton/ha in open field. Data for greenhouse are not available for Azerbaijan.

E) Moghaddam et al. (2011), report of tomato yield in greenhouse 5.52 times higher than in open field in Iran with similar climatic conditions. Comparing the projected yield 485 ton to the open field yield in Israel (81 ton/ha (2008-2010) - FAOSTAT 2012) gives the close ratio 5.99.

F) Kochish et al. (2010). In this study, FCR for 5 breeding crosses of broilers raised in different production systems in one of the Russian large industrial farms are calculated. They belong to the range [1.9, 2.2] when the average FCR equals 2.03.

G) Calculated by data from FAOSTAT (2012), 2008-2010. Includes estimates both for enterprises and individual (less productive) farms.

H) Chudleigh and Simpson (2010). The base scenario in this evaluation of investment assumes the yield 148 kg oil of tea tree per hectare, FCR equals 4.2 for pigs in Belarus (2009).

I) Resolution 568 of the Belarus Government, http://www.government.by/ru/search-solutions/. FCR equals 4.2 for pigs in Belarus (2009).

J) Amuragrocenter, the largest soybean processing plant in the Far East FD, http://amuragro.ru/ (in Russian). Climatic conditions (humidity of beans) and non-genetically modified varieties of soy are similar to those in the project. The oil yield in this extraction plant equals 15.7% (2010).

K) The agro-industrial association Belptizeprom, http://www.agrobel.by/ru/node/23258 (in Russian). The lay rate in industrial poultry farmsin Belarus reached 303 in 2009.

Additional data are available from the author on request.

and genetic materials; oil content; fruit and vegetable yields; productivity; and animal survival rate.

To cope with uncertainty in assumptions and data, the following scenarios were used in business planning and cash flow forecasting:

- Changes in farm integration. E.g.: a) the parent flock against purchasing eggs for a hatchery; b) construction of a meat processing plant and sale of meat products against the sale of chickens and savings in the investment in the plant;
  c) construction of feed mill against purchasing feed mixes.
- Decisions to move to the next phase of the investment. E.g.: investment in capacities for deep-processed products in a soy plant.
- Changes in financing terms and institutional environment. E.g.: changes in interest, loan repayment and grace periods offered by banks; discounts in taxes and customs; and price subsidies.

The spreadsheet bio-economic model for business planning enables a simulation of economic output from changing biological input. The case of an integrated turkey farm was described by Yom Din et al. (2010). In this study, the changing parameters of bird survival rate, egg hatchability, carcass to live animal weight ratio, and eggs layer productivity were used to simulate the internal rate of return for the enterprise.

Project evaluation and ranking

During the evaluation phase of a project, a business plan has to be prepared for consideration by decision makers. If the project is to participate in a tender held under the auspices of a regional government in Russia, it has to meet the conditions defined by government Decree 1470 (1997). The plan also must be consistent with criteria from the investing bank. In Russia, this may be Rosselkhozbank, the largest bank working in the agricultural sector, which published its "Toolkit for Developing a Project Business Plan" in 2007. The financial section of a business plan must include a forecast of economic indicators for the company which will be established in order to implement the project.

In government and bank documents, the recommended criteria for project performance and return on investment are given. For example in 2005, the government of Belarus released Decision 158, which proposed four criteria to evaluate investment project efficiency. These efficiency criteria were based on cash flows (usually, discounted): net income, return on investment, internal rate of return, and payback period.

The Russian government, in the above-mentioned Decree 1470, recommended a payback period along with two other possible criteria:

- A discounted cash flow that takes into account governmental subsidies for the project, along with income and other taxes that the company will pay from earnings (the budget effect of the project);
- A break-even point, where a decline in production leads to zero profits.

In CIS countries, the following criteria are common and are used in this study to evaluate the 14 projects under analysis.

a) The criterion of return on investment (ROI) is defined as the operating profit, divided by the cost of investment. For the purposes of this study, the *operating profit* was calculated as earnings before interest and taxes (EBIT) – the difference between revenue and expense from the normal agro-industrial project activities, not including the effects of interest and taxes. The performance criterion: projects with a high ROI are preferable.

b) The criterion of payback period (PP) is defined as the period of time, in years, required to recover the project investment cost according to the net cash flow (NCF) – as in this study – or discounted NCF. The criterion: projects with a shorter PP are preferable.

c) Internal rate of return (IRR) is the rate of discount that turns the accumulated NCF of the project into zero. The criterion: projects with a higher IRR are preferable. The first two criteria are based on an accounting approach, in which the same importance is given to NCF values obtained at different times. The third criterion is based on an economic approach, in which NCF values receive different weights at different times (discounting). For sectors of agriculture that have a slow initial establishment period for biological reasons – for example the time it takes to establish a herd, or initial growth on a plantation – the assessment period can last up to 7 years or more.

d) To evaluate sensitivity of profit to risk, the following index was used: the sensitivity of project profit, as a percentage, to the simultaneous deterioration of the basic production and market parameters by one percent (see the next section for details). The criterion: projects with lower sensitivity are preferred.

Decision-makers attach varying importance to the project evaluation criteria discussed above. To examine the consistency of project evaluation as calculated by different criteria, the following method was used: all of the analyzed projects were ranked by each of the relevant four criteria. Then the first three sequences of ranks (the performance criteria – ROI, PP, IRR) were examined for internal consistency using Cronbach's alpha index. This statistical index (Cronbach 1951) is widely used, in particular to examine project performance consistency (Khang and Moe 2008). In addition, for each sequence of rank pairs, the significance of the correlation coefficient was estimated.

The value of Cronbach's alpha and the significance of the correlation coefficients enabled a conclusion on performance criteria consistency to be drawn. Finally, the significance of correlation coefficients between ranking by performance measures and by profit sensitivity to risk was estimated.

Profit sensitivity to risk in the cost-volume-profit model
The cost-volume-profit model is based on the following assumptions:

- Agro-industrial enterprise produces and sells a single product (this restriction will be removed later);

- Income and expenses are represented by linear functions of model parameters;
- These parameters – variable and fixed costs, product price – are known, non-random values.

This widely used economic tool is analyzed in the economic literature (Magee 1975; Guidry et al. 1998). In addition to the analysis of management decisions and their financial implications, the opportunity to evaluate profit sensitivity to changes in model parameters is of importance (Kee 2007).

Operational profit P of an enterprise ('project profit') is defined as follows:

$$P = (SP - VC) \cdot Q - FC,$$

where SP is a price of a unit of production,

VC is variable costs per unit of production,

FC is fixed costs of the enterprise, and

Q is sales volume.

A relative change (sensitivity) of profit P is noted as $R_P = \frac{\Delta P}{P}$, where ΔP is a small change in profit. Similarly, a relative change of sales volume is noted as R_Q, of prices as R_{SP}, of variable costs as R_{VC}, and of fixed costs as R_{FC}.

The following formula for relative change in profit under simultaneous change in the four model parameters was derived by Milanovic et al. (2010):

$$R_P = \frac{SP \cdot Q}{P} \cdot (R_{SP} + R_Q + R_{SP} \cdot R_Q) - \frac{VC \cdot Q}{P} \cdot (R_{VC} + R_Q + R_{VC} \cdot R_Q) - \frac{FC}{P} \cdot R_{FC}.$$

In the special case when all model production and market parameters deteriorate by one percent (such deterioration is called "risk" in this article):

$$R_{SP} = R_Q = -1\%, R_{VC} = R_{FC} = 1\%,$$

the approximate formula can be written as follows:

$$R_P = \frac{SP \cdot Q}{P} \cdot (-0.01 - 0.01 + 0.01 \cdot 0.01) - \frac{VC \cdot Q}{P} \cdot (0.01 - 0.01 - 0.01 \cdot 0.01) - \frac{FC}{P} \cdot 0.01 \approx$$

$$\approx \left(-\frac{SP - VC}{P} Q + \frac{-SP \cdot Q - VC \cdot Q - FC}{P} \right) \cdot 0.01 =$$

$$= \left(-\frac{SP - VC}{P} Q + \frac{(SP - VC)Q - FC}{P} - \frac{2SP \cdot Q}{P} \right)\%$$

The expression $\frac{SP-VC}{P} Q$ is the degree of operational leverage DOL, which shows the relative variation in operational profit when sales volume changes by one percent. The DOL is a measure of enterprise business risk (McDaniel 1984). The expression $P/(SP \cdot \backslash, Q)$ shows profitability to sales, and is noted $P_\%$.

Finally, the following expression of profit sensitivity to risk is obtained:

$$R_P = (1 - 2/P_\% - DOL)\%. \tag{1}$$

That is, when cost-volume-profit model parameters deteriorate by one percent, the higher the degree of operating leverage, the higher is decrease in profit, and the higher the profitability, the smaller is decrease in profit.

In the case of n products, the following formula (Milanovic et al. 2010) is used:

$$R_P = \sum_{i=1}^{n} \left(\frac{SP_i \cdot Q_i}{P} \cdot (R_{SP_i} + R_{Q_i} + R_{SP_i} \cdot R_{Q_i}) - \frac{VC_i \cdot Q_i}{P} \cdot (R_{VC_i} + R_{Q_i} + R_{VC_i} \cdot R_{Q_i}) \right) - \frac{FC}{P} \cdot R_{FC},$$

and in a similar way, the approximate formula for profit sensitivity to risk is obtained as follows:

$$R_P = \frac{n}{100} - \sum_{i=1}^{n} (2/P_{\%i} + DOL_i)\%. \tag{2}$$

For every product i, the degree of operating leverage and profitability have the same meaning as in the formula for a single product.

Using the formula (1), elasticity of profit sensitivity to risk is calculated as follows:

$$\varepsilon_{R_P/P_\%} = \frac{2}{100P_\% \cdot R_P} \quad \text{for elasticity to profitability,} \tag{3}$$

$$\varepsilon_{R_P/DOL} = -\frac{P_\%}{100R_P} \quad \text{for elasticity to the degree of operational leverage,} \tag{4}$$

and for the ratio between them:

$$\varepsilon_{R_P/P_\%}/\varepsilon_{R_P/DOL} = -2/P_\%^2. \tag{5}$$

Thus, for projects with reasonable profitability – less than 140% as it follows from (5) – elasticity to profitability is greater in absolute value than elasticity to the degree of operating leverage.

Results

The degree of operating leverage (DOL) and its calculation data are presented in Table 4, with the projects listed in ascending order of DOL. Projects based on crop production are at the bottom of the table, reflecting the high proportion of fixed costs in greenhouses, extraction plants, and perennial plant plantation enterprises. In contrast, an Azerbaijani fruit project is in first place due to the minimal DOL. This project is characterized by a high proportion of variable costs stemming from

Table 4 Project financial indices in ascending order of degree of operational leverage

№	Project	Unit	Quantity	Price, $	Variable costs, US$/unit	Fixed costs, US$ million	Operating profit, US$ million	Degree of operating leverage
3	Fruit, Azerbaijan	ton	2,080	1,058	196	0.3	1.5	1.17
12	Soy, Russia, Far East FD							
	oil	ton	18,240	1,172	761			
	proteins	ton	54,380	1,470	955			
	Total project					6.0	29.5	1.20
10	Turkeys, Russia, Central FD							
	Meat	ton	4,578	4,090	1,332			
	Meat products	ton	5,174	8,165	1,708			
	Total project					11.2	34.9	1.32
9	Pig farm, Belarus	ton	2,751	5,432	540	3.5	10.0	1.35
2	Turkeys, Russia, Privolzhsky FD	ton	13,062	5,120	1,466	12.5	35.2	1.36
5	Broilers, Russia, Privolzhsky FD	ton	21,939	2,698	1,106	9.6	25.3	1.38
13	Eggs, Belarus							
	Eggs	thous. eggs	233,000	78	46			
	Melange	ton	1,800	1,553	869			
	Total project					2.4	6.3	1.39
11	Turkeys, Belarus							
	Meat	ton	3,129	4,168	1,216			
	Meat products	ton	6,068	6,100	1,527			
	Total project					10.5	26.5	1.40
7	Eggs, Romania	thous. eggs	257,000	97.0	44.9	4.0	9.4	1.43
6	Broilers, Kazakhstan	ton	15,031	3,277	1,424	9.7	18.2	1.53
4	Vegetable greenhouses, Azerbaijan	ton	2,500	1,240	80	1.0	1.9	1.54

Table 4 Project financial indices in ascending order of degree of operational leverage *(Continued)*

8	Orchard + extraction	ton	68	22,500	515	0.6	0.9	1.65
14	Milk farm, Russia, North-West FD							
	Milk	ton	22,000	500	136			
	Cattle	ton	540	4,150	645			
	Total project					4.3	5.6	1.78
1	Eggs, Russia, Central FD	thous. eggs	397,000	61.7	36.4	6.1	4.0	2.51

laboratory fruit analysis, long-term storage requirements in a controlled atmosphere, electronic sorting of fruit to determine quality, and packaging requirements for wholesale and retail trade. The high DOL found in project number 1 (eggs) is due to the low profit of the poultry farm under consideration, which operated near the breakeven point. This was caused by lower prices for eggs in Russia until the mid-2000s, while prices increased for resources consumed by this industry over the same period.

The projects were ranked based on profitability data (Table 5) using the procedure described above. Cronbach's alpha index was calculated on the three columns of this table related to performance measures, which received a high value of 0.821. Usually, in economic studies an alpha value of more than 0.7 is considered acceptable to reach a conclusion on internal consistency, which in our case is ranking projects by different measures. Thus it can be concluded that the project evaluation on the basis of the chosen performance measures is consistent (Table 6).

Analysis of correlation coefficients calculated for each of the three possible pairs of ranks lead to a similar conclusion. This analysis revealed the high significance of the correlation coefficients (1%–5%).

At the same time, the correlation coefficients between each ranking by performance criteria and ranking by profit sensitivity are close to zero (Table 7).

Values of elasticity of profit sensitivity to risk were calculated for every project using formulae (3) and (4). They returned the following values: elasticity to profitability from 0.83–0.97% and elasticity to DOL from 0.01-0.23%. The projects estimated profitability ranged from 16–70% (Table 5). For every project, the ratio between elasticity values to profitability and to DOL was much greater than unity (Table 8). This means that in terms of elasticity, profit sensitivity to risk was influenced by the project profitability to a much greater degree than by DOL, in accordance with the remark after (5).

Table 5 Project profitability: the profitability measures calculated for the studied projects

№	Project, country, Federal District	Return on investment	IRR	Payback period, years	Profitability to sales
1	Eggs, Russia, Central	6.2%	3.0%	12.5	16%
2	Turkeys, Russia, Privolzhsky	27.1%	26.2%	3.7	53%
3	Fruit, Azerbaijan	18.7%	19.8%	9.3	70%
4	Vegetable greenhouses, Azerbaijan	17.2%	26.6%	4.3	61%
5	Broilers, Russia, Privolzhsky	20.5%	23.5%	4.5	43%
6	Broilers, Kazakhstan	22.1%	30.0%	4.8	37%
7	Eggs, Romania	21.9%	32.1%	3.5	38%
8	Orchard + extraction	38.3%	24.0%	8.0	59%
9	Pig farm, Belarus	19.8%	26.6%	7.5	67%
10	Turkeys, Russia, Central	19.6%	27.5%	5.3	57%
11	Turkeys, Belarus	18.9%	26.6%	7.2	53%
12	Soy, Russia, Far East	26.82%	39%	4.0	29%
13	Eggs, Belarus	21.1%	20.6%	6.0	30%
14	Milk farm, Russia, North-West	13.0%	8.5%	9.0	32%

Table 6 Project ranking by performance measures and profit sensitivity to risk

№	Project	Measures of performance*			Profit sensitivity
		Return on investment	IRR	Payback period	
1	Eggs, Russia, Central FD	14	14	14	14
2	Turkeys, Russia, Privolzhsky FD	2	8	2	6
3	Fruit, Azerbaijan	10	12	13	1
4	Vegetable greenhouses, Azerbaijan	12	6	4	4
5	Broilers, Russia, Privolzhsky FD	7	10	5	8
6	Broilers, Kazakhstan	4	3	6	10
7	Eggs, Romania	5	2	1	9
8	Orchard + extraction	1	9	11	5
9	Pig farm, Belarus	8	7	10	2
10	Turkeys, Russia, Central FD	9	4	7	3
11	Turkeys, Belarus	4	3	6	10
12	Soy, Russia, Far East FD	3	1	3	12
13	Eggs, Belarus	6	11	8	11
14	Milk farm, Russia, North-West FD	13	13	12	13

* Cronbach's alpha for measures of performance.

Discussion

The agro-industrial project market in the CIS and EE countries studied is favorable for companies and developers. In terms of investment, the volume for the five largest countries (Russia, Ukraine, Belarus, Kazakhstan, and Romania) comes close to US $18 billion per annum. The upward trend in prices for crop and livestock production continues in the last years. The high proportion of food expenditure in the household budget, the growing demand for food, and the potential for a substantial increase in food exports indicate that it is very likely that the present demand for agro-industrial projects will not decrease in the near future. The projects are supported by regional and central governments by subsidies of interest on bank loans, reduced taxes, and subsidies for less profitable enterprises (milk, eggs). A high degree of vertical integration and use of the latest technological solutions contribute to the success of the projects analyzed.

Project evaluation by various performance measures is a prerequisite for success in this market. For the projects studied, different performance criteria have been shown to provide ranking consistency. This is a good basis for successful integration of the results, which is important for several decision-makers. The economic literature offers various methods for the integration of investment project evaluations by different performance criteria (Chong 2008; Subramanian and Ramanathan 2012).

Table 7 Correlation coefficients between ranking by performance measures and by profit sensitivity to risk

	Return on investment	IRR	Payback period	Profit sensitivity
Return on investment	1			
IRR	0.54*	1		
Payback period	0.53*	0.74**	1	
Profit sensitivity	0.04	0.07	−0.02	1

* Significance at 5%.
** Significance at 1%.

Table 8 Elasticity calculation of profit sensitivity to risk

№	Project, country, Federal District	$P_{\%}$ - profitability to sales, from Table 5	DOL - degree of operating leverage, from Table 4	R_P - profit sensitivity to risk, from equation (2)	Elasticity to profitability, from equation (3)	Elasticity to DOL, from equation (4)
1	Eggs, Russia, Central	16%	2.51	−13.75%	−0.89%	0.01%
2	Turkeys, Russia, Privolzhsky	53%	1.36	−4.16%	−0.91%	0.13%
3	Fruit, Azerbaijan	70%	1.17	−3.03%	−0.95%	0.23%
4	Vegetable greenhouses, Azerbaijan	61%	1.54	−3.84%	−0.86%	0.16%
5	Broilers, Russia, Privolzhsky	43%	1.38	−5.06%	−0.92%	0.09%
6	Broilers, Kazakhstan	37%	1.53	−5.95%	−0.91%	0.06%
7	Eggs, Romania	38%	1.43	−5.75%	−0.93%	0.07%
8	Orchard + extraction	59%	1.65	−4.04%	−0.84%	0.15%
9	Pig farm, Belarus	67%	1.35	−3.34%	−0.90%	0.20%
10	Turkeys, Russia, Central	57%	1.32	−3.82%	−0.92%	0.15%
11	Turkeys, Belarus	53%	1.40	−4.18%	−0.90%	0.13%
12	Soy, Russia, Far East	29%	1.20	−7.07%	−0.97%	0.04%
13	Eggs, Belarus	30%	1.39	−7.07%	−0.95%	0.04%
14	Milk farm, Russia, North-West	32%	1.78	−7.44%	−0.83%	0.04%

Market changes and different levels of integration have been taken into account when comparing projects from the same industry for different years. For instance, for projects 2 (2006) and 10 (2011), both in the turkey industry, investment for the latter project is greater at 29% (Table 2), though production volume is lower (23%) and the additional cost of the meat products plant (Table 3) accounts for only 4% of the total investment. This increased investment is caused by the rise in construction costs and equipment in Russia in the period 2006–2011, and explains low indices of return on investment and payback period for project. At the same time, most of the sales for this more integrated project come from meat products, which are more profitable than the chilled meat that made up a large part of the sales volume from project 2. This explains why profitability and internal profitability (IRR) are somewhat higher in project 10.

In the projects studied no correlation between their ranking by performance criteria and by profit sensitivity to risk was shown. For example, the Azerbaijani fruit enterprise was ranked first in low profit sensitivity to risk. This project employed long-term fruit storage technology and was characterized by a high proportion of variable costs and by a long period between the initial investment to a full commercial crop. The first feature explains the low degree of operating leverage and, given the high operating profitability (70%), the low profit sensitivity to risk. The second feature lead to a negative cash flow during the first 4–5 years of the project. This leads accordingly to low IRR and a long payback period (Table 6). This example illustrates a possible scenario for agro-industrial projects in which ranking by performance evaluation can be inconsistent with ranking by profit sensitivity to risk evaluation.

Profit sensitivity to risk for the surveyed projects is high, with changes from 3-14% in response to a simultaneous one percent deterioration of production and market parameters (Table 8). However, the risk of parameter deterioration can be lower than that of traditional agriculture due to the industrial nature of the technology used, significant amount of storage capacity for raw materials, large sales volume from products processed from raw materials, ability to create financial reserves, and the ongoing analysis of animal condition and of raw materials and product quality. The projects where the main technological characteristics are close to the host country's typical values, and do not show significant advantages from this criterion (projects 1, 5, 6, 12, 13 in Table 3) are the most sensitive projects, with ranks ranging from 8–14 (Table 6).

Conclusions

From the perspective of the agro-industrial projects developers, the market in the studied countries of the CIS and EE is large and varies by region and industry. The market is favorable to developers due to external (world price growth) and internal (positive changes in the institutional environment due to food security concerns) factors. The decision-makers in this market – the *region*, the *bank*, and *the initiator* – use different evaluation criteria.

A method for project evaluation using the criterion of profit sensitivity to risk is proposed. The approximate formulas for profit sensitivity to risk (when basic production and market assumptions change simultaneously) and its elasticity are derived, based on the cost-volume-profit model. The formulae enable reduced calculations to explain profit sensitivity and elasticity using the usual indicators of business planning: operational profitability and degree of operating leverage.

In order to examine the project ranking consistency by different criteria, Cronbach's alpha and correlation coefficients were used. For the studied projects, their ranks by performance measures are consistent with each other, but are not correlated with their ranking by profit sensitivity to risk. The positive result from this inconsistency is that evaluating projects by profit sensitivity to risk may provide useful, additional information to decision-makers.

The main limitation of the derived approximate formulae for profit sensitivity to risk analysis is found in the assumption that the used cost-volume-profit model is linear to its four main parameters. Analysis with the removal of this assumption is an area for future research.

Competing interests

The authors declares that he has no competing interests.

Acknowledgements

The author gratefully acknowledges experts and colleagues from the following Israeli companies: AgriGo, Agro Technology, AgroTop, as well as agronomists Ilan Sela, Dror Dagan, Dr. Zinaida Zugman, and Professor Roman Sheinberger for their professional advice and insight in the preparation of this article.

References

BelStat (2012) National Statistical Committee of the Republic of Belarus. http://www.belstat.gov.by/homep/en/main.html. Accessed 19 Mar 2013

Ben-Zion U, Zolotoy L, Yom Din G (2005) Skewness Aversion, Output Uncertainty and Hedging in a Futures Market. Available via SSRN: http://ssrn.com/abstract=673344 Accessed 19 Mar 2013

Berkowitz D, DeJong D (2011) Growth in post-soviet Russia: a tale of two transitions. J Econ Behav Organ 79(1–2):133–143

Bevzeluk A (2008) Metodi ocenki investicionnykh proektov. Bankajski vesnik 19(lipen):12–18, (Банкаўскі веснік 19 (лінень)). In Russian

Bojnec Š, Fertő I (2012) Does EU enlargement increase agro-food export duration? World Econ 35:609–631

Chien YI, Cunningham WHJ (2000) Incorporating production planning in business planning: a linked spreadsheet approach. Prod Plan Control 11(3):299–307

Chudleigh P, Simpson S (2010) Economic Evaluation of Investment in the Tea Tree Oil R&D Program. RIRDC Publication No 10/212: ISBN 978-1-74254-177-8, ISSN 1440–6845

Chong G (2008) How to Appraise Investment Projects. J Corp Accounting & Finance 19(2):59–64

Commodity Markets (2012) The World Bank. http://www.worldbank.org/prospects/commodities. Accessed 19 Mar 2013

Cronbach LJ (1951) Coefficient Alpha and the Internal Structure of Tests. Psychometrika 16:297–334

EuroStat (2012a) Consumption expenditure of private households. http://epp.eurostat.ec.europa.eu/portal/page/portal/eurostat/home. Accessed 19 Mar 2013

EuroStat (2012b) National Accounts. http://epp.eurostat.ec.europa.eu/portal/page/portal/national_accounts/data/database. Accessed 19 Mar 2013

FAOSTAT (2012) Data collection of the Statistics Division of the Food and Agriculture Organization of the United Nations. http://faostat.fao.org/. Accessed 19 Mar 2013

Goldman MI (2003) The piratization of Russia: Russian reform goes awry. Routledge, London

Guidry F, Horrigan JO, Craycraft C (1998) CVP analysis: a new look. J Managerial Issues 10(1):74–85

Hockmann H, Gataulina E, Hahlbrock K (2011) Risk, technical efficiency and market transaction costs in different organizational forms: evidence from the oblast Tatarstan. No 114510, 51st Annual Conference, Halle, Germany, German Association of Agricultural Economists (GEWISOLA). http://ageconsearch.umn.edu/bitstream/114510/2/Hockmann_et_al.pdf. Accessed 19 Mar 2013

Isakov IY (2011) Application of the state financial institutions of development for activization of investment process. Polythematic power electronic scientific journal of the Kuban State Agrarian University, №67(67) (in Russian), http://ej.kubagro.ru/2011/03/pdf/14.pdf. Accessed 15 April 2013

Kalugina ZI (2011) Vector of postcrisis development of rural Russia. Regional Research of Russia 1(2):149–156

KazStat (2012) Statistics Agency of the Republic of Kazakhstan. http://stat.kz/Pages/default.aspx. Accessed 19 Mar 2013

Kee R (2007) Cost-volume-profit analysis incorporating the cost of capital. J Managerial Issues 19(4):478–493

Khang DB, Moe TL (2008) Success criteria and factors for international development projects: a lifecycle based framework. Project Management J 39(1):72–84

Kochish I, Fedkina T, Kovinko V (2010) Genotype, environment and broilers' performance. Animal Husbandry of Russia, 9:11–12 (in Russian). http://zzr.ru/archives/2010/09/09-2010_02.pdf. Accessed 19 Mar 2013

Ksenofontov MY, Gromova NA, Polzikov DA (2012) Forecast-analytical studies on the priorities of agrifood policy. Studies on Russian Economic Development 23(2):115–127

Kuehne G, Nicholson C, Robertson M, Llewellyn R, McDonald C (2012) Engaging project proponents in R&D evaluation using bio-economic and socio-economic tools. Agricultural Systems, 108:94-103. http://www.sciencedirect.com/science/article/pii/S0308521X1200025X. Accessed 14 Apr 2013

Liefert WM, Liefert O (2012) Russian agriculture during transition: performance, global impact, and outlook. Applied Economic Perspectives and Policy 34(1):37–75

Lioubimtseva E, Henebry GM (2012) Grain production trends in Russia, Ukraine and Kazakhstan: new opportunities in an increasingly unstable world? Frontiers of Earth Science 6(2):157–166

Magee RP (1975) Cost-volume-profit analysis, uncertainty and capital market equilibrium. J Accounting Res 13(2):257–266

Mansurov PE (2011) Kriterii ocenki konkurentosposobnosti agropromishlennogo predpriyatiya. Vestnik Rossiyskoy Akademii selskohozyaystvennyh nauk 3:22–24 (in Russian). http://elibrary.ru/contents.asp?issueid=937344&selid= 16378021. Accessed 19 Mar 2013

McDaniel WR (1984) Operating leverage and operating risk. J Bus Finance & Accounting 11:113–125

Milanovic D, Lj MDD, Misita M, Klarin M, Zunjic A (2010) Universal equation for the relative change in profit of manufacturing company. Prod Plan Control 21(8):751–759

Moghaddam PR, Feizi H, Mondani F (2011) Evaluation of tomato production systems in terms of energy use efficiency and economical analysis in Iran. Not Sci Biol 3(4):58–65

Otiman PI (2012) Romania's present agrarian structure: a great (and Unsolved) social and economic problem of our country. Agr Econ & Rural Development 9(1):3–24

Parfenova EN (2009) Problemi metodiki ocenki regionalnykh investicionnykh proektov. Nauchnye vedomosti BelGU. Ser History, Politology, Economics 12(1):22–27 (in Russian)

Petrick M, Wandel J, Karsten K (2012) Economic & social impacts of recent agro-investment in Kazakhstan's grain region. In: Annual World Bank Conference on Land and Poverty. http://www.landandpoverty.com/agenda/pdfs/paper/ petrick_full_paper.pdf. Accessed 19 Mar 2013

Romania Stat (2012) National Institute of Statistics, Romania. http://www.insse.ro/cms/rw/pages/index.en.do. Accessed 19 Mar 2013

RosStat (2012) Federal State Statistics Service. http://www.gks.ru/wps/wcm/connect/rosstat/rosstatsite/main/. Accessed 19 Mar 2013

Saubanov KR (2010) Konkurentosposobnost regionalnogo selskogo khozyaistva vprivolzhskom federalnom okruge: otsenka i puti povisheniya. PhD thesis (in Russian). http://www.dissercat.com/content/konkurentosposobnost- regionalnogo-selskogo-khozyaistva-v-privolzhskom-federalnom-okruge-otse. Accessed 15 April 2013

Sauer J, Gorton M, White J (2012) Marketing, cooperatives and price heterogeneity: evidence from the CIS dairy sector. Agr Econ 43(2):165–177

Shepotylo O (2012) Export diversification across countries and products: do Eastern European (EE) and Commonwealth of Independent States (CIS) countries diversify enough? J Int Trade & Econ Development 21:1–34

The State Program for Development of Agriculture and Regulation of Agricultural Commodities Markets in 2013–2020 (2012) The Ministry of Agriculture of the Russian Federation (in Russian). http://government.ru/gov/results/19885/. Accessed 19 Mar 2013

Strashko IV (2010) Business-planirovanie investicionnykh proektov v APK: conceptualnyj podhod. Problemi sovremennoy ekonomiki 1(33):414–417 (in Russian)

Subramanian N, Ramanathan R (2012) A review of applications of analytic hierarchy process in operations management. Int J Prod Econ 138(2):215–241

Torok A, Jambor A (2012) Changes in agri-food trade of the new member states since EU accession - a quantitative approach. http://ideas.repec.org/p/ags/iaae12/125140.html. Accessed 19 Mar 2013

Turkey of Stavropol region (2012) Selective Breeding Genetic Center. http://indeikastav.ru/en/incubation12.html. Accessed 19 Mar 2013

UkrStat (2012) State Statistics Service of Ukraine. http://ukrstat.gov.ua/. Accessed 19 Mar 2013

Vasina NV (2012) Modelirovanie finansovogo sostoyaniya selskohozyajstvennih organizacij pri ozenke ih kreditosposobnosti., ISBN 978-5-98566-078-4, Omsk (in Russian)

Visser O, Mamanova N, Spoor M (2012) Oligarchs, mega-farms and land reserves: understanding land grabbing in Russia. J Peasant Stud 39(3–4):899–931

Voicilas DM (2011) Investment process in Romania and institutional dysfunctionalities. J Eur Stud & International Relations 14(2):359–370

Voigt P, Hockmann H (2008) Russia's transition process in the light of a rising economy: economic trajectories in Russia's industry and agriculture. Eur J Comparative Econ 5(2):179–195

Wandel J, Pieniadz A, Glauben T (2011) What is success and what is failure of transition? A critical review of two decades of agricultural reform in the Europe and Central Asia region. Post-Communist Econ 23(2):139–162

Wegren SK (2012) Institutional impact and agricultural change in Russia. J Eurasian Studies 3(2):193–202

Wise TA, Murphy S (2012) Resolving the Food Crisis: Assessing Global Policy Reforms Since 2007. Institute for Agriculture and Trade Policy, Global Development and Environmental Institute. http://iatp.org/documents/ resolving-the-food-crisis-assessing-global-policy-reforms-since-2007. Accessed 19 Mar 2013

Yom Din G, Gilad S, Zugman Z (2010) A model for estimating how variability of biological parameters affects economic factors in an integrated turkey farm. Comput Electron Agr 75(1):100–106

Zaharov AN (2006) O perspektivakh razvitiya rossijskogo APK (in Russian). http://ecsocman.hse.ru/data/535/626/1219/ zaxarov.pdf. Accessed 19 Mar 2013

Zhang W, Wilhelm WE (2011) OR/MS decision support models for the specialty crops industry: a literature review. Ann Oper Res 190(1):131–148

5

Bioenergy chain building: a collective action perspective

Luigi Cembalo[1][*], Francesco Caracciolo[1], Giuseppina Migliore[2], Alessia Lombardi[1] and Giorgio Schifani[2]

* Correspondence:
cembalo@unina.it
[1]Department of Agriculture, AgEcon and Policy Group, University of Naples Federico II, Via Università 96, Portici, NA 80055, Italy
Full list of author information is available at the end of the article

Abstract

Depletion of natural resources has become a key issue on the European policy agenda. Bottom-up measures have emerged in several countries with a view to promoting awareness campaigns and environmental sustainability, with the agenda set by individuals who start up collective initiatives at the local level. Such collective action provides an incentive to free-ride on the contribution of others. Social norms and the consequent behavior of individuals involved in collective action assume a key role in ensuring sustainable use of a public good, achieving significant, long-lasting success. The present study aims to ascertain which determinants most affect farmers' willingness to contribute to common resources. The empirical study was conducted in an area in the province of Avellino (southern Italy) most affected by soil erosion problems. The study focused on the willingness of farmers to contribute to the public good through biomass production (Giant Cane). In all, 175 face-to-face questionnaires were administered to farmers in September-November 2013. Schwartz's norm-activation model variables were collected. A Tobit model was implemented in which the dependent variable was the land farmers stated they were willing to cultivate with Giant Cane. Four on five psychological constructs, based on the NAM, proved statistically significant with the expected sign, showing that an altruistic behavioral approach is useful to predict the individual's decision to adopt cooperation norms.

Keyword: Public goods; Norm-activation model; Tobit

Background

Environmental pollution, biodiversity loss, climate change and the inexorable depletion of natural resources have become major issues on the European policy agenda. However, European governments have been very reluctant to adopt radical approaches to solve such problems (Rootes, 2012; Hisschemöller and Sioziou, 2013). In this context, bottom-up measures have emerged in several countries with a view to promoting awareness campaigns and practical action aimed at environmental sustainability (Caracciolo and Lombardi, 2012). By their very nature, such measures are set up by individuals who offer their time and resources to start up collective action initiatives at the local level. This has major implications for the theory of public goods. Indeed, no "rational selfish" individual would be willing to contribute to the production of a public good even when, in a condition of cooperation with other individuals, this contribution would entail a reciprocal benefit (Liebe et al. 2011). This assumption, known as the "zero contribution thesis", has been contradicted by some theoretical and empirical cases (Olson, 1965; Ostrom, 1991; Gibson et al. 2005; Ostrom, 2010; Sayfang 2010). However, the production of a public good entails the non-excludability of individuals from (direct or indirect) use of the good itself, also for individuals who do not contribute to supplying the good. This clearly provides an incentive to rely on the contribution of others and use the good as a free rider. When such a situation occurs, it is referred to as a social dilemma (Olson, 1965; Dawes et al. 1986; Ostrom, 2000). A social dilemma may be defined by a situation where each member of a group gains a better payoff if he/she pursues his/her personal interest but, at the same time, everyone benefits from the fact that all members of that group aim at a common interest.

Better and more sustainable management of common resources is achieved when the rules concerning the use of common resources are defined by involving economic subjects who are in a situation of interdependence (Ostrom, 2000; Ostrom 2010; Ostrom and Walker 2003). The social norms that arise as a result of interactions aim to reduce the problem of free-riding, since they are based on relations of reciprocity and trust (Ostrom, 2010). This assumption is a core idea in the theory of public goods and collective actions. Following this line of reasoning, social/personal norms and the consequent behavior of individuals involved in collective action assume a key role in ensuring sustainable use of such goods, achieving significant, long-lasting success. Such success is assured by the rules put in place in virtue of a reciprocal consensus within the community (Berkes, 1987). This is why, in order to achieve a collective action with high payoffs, it is necessary that individuals share social norms of cooperation. However, individuals show different willingness to start reciprocity and then collective action. More precisely, as illustrated by Ostrom (2000), collective action may be initiated thanks to the presence of conditional cooperators. Therefore, individuals "*who are willing to initiate cooperative action when they estimate others will reciprocate and to repeat these actions as long as a sufficient proportion of the others involved reciprocate*" (Ostrom, 2000 - p. 142). Put differently, when some individuals initiate cooperation, others learn to trust them and are more willing to adopt reciprocity themselves, leading to higher levels of cooperation. Conditional cooperators are thus the source of relatively high levels of contribution in the management of common resources. From here it emerges that social dilemma and trust in other people's cooperation represents the most important determinants that affect collective action.

Several types of collective action have relevance to natural resource management. To the best of our knowledge, scholars have restricted their studies only to long-surviving cooperation systems, like use and governance of forests (Agrawal and Goyal, 2001; Antinori and Bray, 2005; Gibson et al. 2005), irrigation systems (Meinzen-Dick et al. 2002; Fujiie et al. 2005) and food community (Migliore et al. 2014, 2015). The existence of different types of individuals allows the assumption that among economic actors some are able to contribute to the management of common resources. To understand whether conditions to generate collective action exist, dilemma concern and trust in other people's cooperation are not sufficient predictors of such behavior.

The present study aims to ascertain what determinants most affect farmers' willingness to contribute to common resources. The hypothesis underlying the study is that an altruistic behavioral approach, using psychological constructs, could be useful to predict the individual's decision in sharing cooperation norms. Cooperative behavior is more likely when economic actors are aware of the consequences and aware of the responsibility of their action on others, in a context featuring social dilemma and trust in other people. Among public goods, an important role may be played by biomass production in an agro-energy chain.

The empirical study was conducted in an area in the province of Avellino (southern Italy), most affected by soil erosion problems. The main crop in the area is currently wheat. The study focused on the willingness of farmers to contribute to the public good through biomass production. The biomass crop suggested was *Arundo donax* (Giant Reed); it was chosen because of its high biomass productive efficiency, its ability to significantly mitigate soil erosion risk (it is a multi-year crop), and for its capacity to yield an income comparable to wheat. Participants in the proposed, though hypothetical, agro-energy chain were informed of the positive environmental effects of producing the crop in question: increased production of agro-energy means a reduction in pollution emissions (Kotchen and Moore, 2007); and contributes to mitigate soil erosion risk.

About 175 face-to-face questionnaires were administered to farmers in September-November 2013. Schwartz's norm-activation model (NAM) variables were collected (Schwartz, 1977; Schwartz and Howard, 1980). A principal component analysis was performed on NAM variables, allowing identification of five psychological constructs namely: social dilemma, awareness of responsibility, personal norms, trust in other people's cooperation and awareness of consequences. Factorial scores of the above constructs, as well as some farmers' and farm characteristics, were implemented in a Tobit model in which the dependent variable was the amount of land stated by farmers potentially involved in giant reed cultivation. Four out of five psychological constructs proved statistically significant with the expected sign, showing that an altruistic behavioral approach, by using psychological constructs, is useful to predict the individual's decision to adopt cooperation norms.

The paper is organized as follows: in section Norm activation model of pro-social behavior we review the theoretical framework for describing farmers' motivational attitudes toward participation. In section Methods we describe the data and the empirical model, while results of the econometric model are presented in section Results and discussion. Conclusions follow, including some important caveats and limitations.

Norm activation model of pro-social behavior

Environmental depletion problems present an intrinsic contradiction between the optimal choice for a rational individual and the social optimum. The trade-off between individual and collective benefit has often been conceptualized within psychological models of altruistic behavior such as Schwartz's norm-activation model (NAM) (Schwartz, 1977; Schwartz and Howard, 1980). In general terms, Schwartz argued that personal norms are the only direct determinants of pro-social behavior patterns. One important class of pro-social behavior, with broad applicability, is cooperation.

According to Schwartz, personal (or moral) norms influence behavior when actors are aware that certain actions have consequences on others' wellbeing (called awareness of consequences), accepting the responsibility of those actions (awareness of responsibility). Put differently, relationships between personal norms and cooperative behavior is moderated by the awareness of consequences (AC) and by the awareness of responsibilities (AR) of such actions on other people. The NAM proposed by Schwartz generated diverse approaches applied both in social (Hopper and Nielsen 1991) and in environmental contexts (Stern et al. 1999; Joireman et al. 2001; De Groot & Steg, 2009). Among the latter, the NAM was successfully implemented in empirical studies concerning the willingness to pay for environmental protection (Liebe et al. 2010; Guagnano, 2001; Guagnano et al. 1994). In these studies, however, the relationships among NAM's key factors remain fuzzy. Some scholars find that personal norms are the best predictors of environmental behavior (Stern et al. 1999). Others indicate that awareness of consequences on the environment, in agreement with personal norms, can be the main predictors of pro-environmental behavior (Hopper and Nielsen, 1991; Vining and Ebreo, 1992). Others, again, suggest that extra factors be implemented in the NAM (Blamey, 1998; Joireman et al. 2001).

Indeed, it has been shown that social rather than personal norms are related to environmental behavior (Ebreo et al. 2003). Guagnano et al. (1995) in a study on waste recycling found that external conditions, that is, personal costs and presence or otherwise of recycling bins, affect the relationship between the key factors of the NAM and behavior. In line with these findings we believe that other factors could be implemented within the NAM. According to the literature on participation in natural resources management and collective action, the additional factors should be sought after in the concept of social capital. Importantly, there are many interpretations of social capital, but one useful definition was made by Ostrom: "...*the shared knowledge, understandings, norms, roles, and expectations about patterns of interactions that groups of individuals bring to a current activity...*" (Ostrom, 1999:176). In this regard, relations of trust, reciprocity and exchanges, common rules and norms are often viewed as important mechanisms for building social capital assets (Pretty and Ward, 2001).

Nevertheless, in a perspective of collective action the key question is not whether any one individual will contribute, but whether enough individuals will contribute rather than free-ride. Individuals are seen to have contingent strategies or preferences, cooperation being contingent on certain aspects of the choice-situation. For example, cooperation has been found to be more likely when it is perceived that the good will only be provided if every member of the collective action contributes. In other words, cooperation is more likely when it is perceived that collective action will have a desirable outcome. For these reasons the key determinants of willingness to contribute to the public good should include social dilemma and trust in other people's cooperation. More precisely, we assume that pro-social behavior is predicted by five types of determinants: social dilemma, trust in other people, social/personal norms, awareness of consequences and awareness of responsibility. Among these, social/personal norms are related to a "moral obligation to perform or refrain from specific actions" (Schwartz & Howard, 1981), p. 191. The awareness of consequences is defined as whether someone is aware of the positive or negative consequences for others when acting or not acting pro-socially to protect the environment. Finally, awareness of responsibility is described as a condition in which "a person believes he or she can make a useful contribution to the solution of the problem, whith perceived outcome efficacy" (Montada, and Kals, 2000; De Groot and Steg, 2009). However, according to the literature on collective action, it is important to stress that trust in other people occurs when individuals are engaged in an interaction process. Therefore it is possible to suppose that in an early-stage, hypothetical, collective action this variable could be difficult to predict.

Methods

The data collected through the administration of the questionnaire with face-to-face interviews are discussed here in relation to the main questionnaire sections. The aim was to collect information to interpret and understand the choice behavior, and the psychological constructs, of 200 interviewed farmers. Farmers' socio-demographic data were also collected, as well as details on their general farm organization. Finally, farmers' most recent choices in terms of investments and innovation in their farm were collected as well as in terms of participation in formal or informal forms of cooperation or contract.

We first drafted a pilot questionnaire. A final version of the questionnaire was then organized into three main sections: first, socio-demographic; secondly, a detailed description of biomass cultivation, specifically giant reed, the set-up of the bioenergy chain in the study area, the need for collective action from the local farmers; thirdly, the set of questions related to the NAM. As for the latter, a seven-point Likert scale was used to help respondents express their level of agreement between the statements provided and their own motivations.

Administration of the face-to-face survey in the case-study area provided 200 questionnaires from the randomly selected sample. Descriptive statistics were collected on the 200 farmers in question although only 171 were used in the model due to data missing in the NAM section of the questionnaire. Based on the data collected by the first section of the questionnaire (summarized in Table 1), in our sample the entrepreneurs are mostly male (63%), aged on average 43, full-time farmers and owners of the land they cultivate.

The sample farms cover an area of 4092 ha, of which 3706 ha are cultivated. Farm size is quite variable in the sample, ranging from a minimum of 1.5 ha to a maximum of 340 ha. However, the most frequent size (sample mode) is 20 ha. Farms are generally

Table 1 Farmers' SDs and farm characteristics

	Description	Mean	Std.dev	Min	Max
Gender	1 if male; 0 female	0.61	N.A.	0	1
Age		43.17	11.76	18	80
Farm size	Total area of sample farms (ha)	20.46	26.66	1.5	340
Total cultivated area	Total cultivated area of sample farms (ha)	18.59	23.52	1	300
Full-time farmer	1 if full-time; 0 otherwise	0.77	N.A.	0	1
Individual farmers	1 if individual; 0 otherwise	0.96	N.A.	0	1
Number of plots		6		1	30
Land ownership					
	Property	0.72	N.A.	0	1
	Rent	0.26	N.A.	0	1
	Other	0.02	N.A.	0	1
Crops					
	Arable crops	16.70	21.91	0.5	280
	Fruit trees	0.5955	1.14	0	9

N.A: not applicable.

fragmented into several plots, varying from a minimum of 1 up to 30 plots, with an average number of six per farm. Arable crops are clearly predominant in the sample farms (93% of the total cultivated land). Only a very small share of the land is left for olive trees, industrial crops and grapevines.

Erosion and landslides, along with steep slopes, are the most important causes of land abandonment in the sample farms. Indeed, for the above reasons in 60 farms plots accounting for 178 ha are no longer cultivated.

Entrepreneurs' attitudes to change and openness to innovation were explored by investigating the investments and innovations introduced into their farms in the last five years, as summarized in Table 2. In all, 70% of entrepreneurs had invested in new equipment or new processing systems or had bought more land during the past five years. Considerably fewer (46%) had introduced some innovation in cropping systems, techniques and farm organization.

Not surprisingly, the younger entrepreneurs are more open to new investments and innovations: the mean age of the "innovators" (41) is lower than the mean age of the whole sample (43) and the most frequent value (mode) is 38. Cooperation and collective contracts are not particularly common in the sample. Although more than half of the sample (56%) are aware of the presence of some forms of farmer cooperation and contracts in the area, only 32% of the sample entrepreneurs have participated, focusing mainly on the marketing side to control uncertainty, to secure a minimum price and access to markets (Table 3).

Farmers who participate in different forms of cooperation are generally younger (40 years old) than the sample average (43), and farm size is larger (25.58 ha) than the average size of the sample (20.46 ha). The most common form of contract they are involved with (68%) is informal and one-year long.

The last section of the questionnaire served to collect NAM (or psychological constructs) variables. In Table 4 the full statements presented to collect NAM variables and some descriptive statistics are reported.

A crucial aspect in the analysis regards the elicitation of the key response variable of interest: an open-ended question was used to elicit farmers' intentions to participate in the bioenergy supply chain. The latter was measured by means of the stated intention

Table 2 Investments and innovation in the past five years*

	Absolute freq.	Relative freq.
Investments in the last 5 years:	139	0.70
Machines	119	0.60
New constructions	66	0.33
Processing and packaging	6	0.03
Marketing	11	0.06
Land acquisition	83	0.42
Other	1	0.01
Innovations in the last 5 years:	91	0.46
Cropping system changes	35	0.18
Cropping technique changes	9	0.05
Organization changes	64	0.32

*The number of investments and innovations is greater than the number of farms as some farms have adopted more than one.

Table 3 Presence of forms of cooperation in the area and participation

	Absolute freq.	Relative freq.
Forms of cooperation in the area:	111	0.56
Informal contracts	66	0.33
Cooperatives and trademarks	40	0.20
Supply chain contracts and guaranteed minimum price	2	0.01
Participation in cooperation activities:	64	0.32
Cooperation on the production side*	16	0.08
Cooperation on the marketing side**	48	0.24

*Technical assistance, supply of production inputs and raw materials, transfer of technological innovations.
**Product marketing, minimum price guarantee, less uncertainty in product allocation, access to markets.

of farmers to allocate a share of their land (measured in hectares) for cultivating giant reed (AB). This response could be assumed as a linear, additive and separable function of the h-farmer's observed individual attributes (IA_h), of the structural characteristics (FC_h) of the farms owned, and of the h-farmer's psychological constructs, or NAM variables (PC_h). Our research focused on identifying which of the above factors may

Table 4 Statements to collect NAM (psychological constructs) variables and the main descriptive statistics - (Obs. 171)

Statement	Variable code	Mean	St.dev	Min	Max
Social Dilemma (−)					
If individually I do something for the environment it will not change anything	NAM1	3.15	2.231	1	7
If other local farmers do not participate in biomass agroenergy, I will not be willing to contribute to this supply chain either.	NAM2	2.64	1.927	1	7
If I do something for the environment, I am ingenuous, because I will always suffer the environmental consequences of damaging action committed by others	NAM3	2.72	1.959	1	7
Awareness of responsibility (+)					
I see myself as a possible activist in the production of biomass (*Arundo donax*)	NAM4	4.56	1.610	1	7
I would like, in the future, to contribute actively to the production of energy from renewable and more environmentally sustainable resources.	NAM5	5.99	1.156	1	7
In the coming weeks I will improve my knowledge concerning the possibility of cultivating *Arundo donax* for biomass production	NAM6	5.29	1.348	1	7
Personal/social norms (+)					
Participating in a collective action for producing biomass is for me important in order to contribute to helping the environment	NAM7	6.11	1.119	1	7
Contributing to the stewardship of the environment is a moral obligation	NAM8	5.85	1.427	1	7
Trust in other people's cooperation (−)					
I believe also other farmers in my area would be willing to start collective action for the production of biomass, obtaining clean energy and helping the environment	NAM9	5.08	1.493	1	7
My family and friends would be proud if I contributed to the production of energy from renewable resources through biomass production	NAM10	5.44	1.406	1	7
Awareness of consequences (+)					
The best way to solve environmental issues is to act collectively	NAM11	6.68	0.795	1	7
Improvement in environmental conditions can only be achieved through collective action	NAM12	6.01	1.581	1	7

Expected signs in parentheses.

affect and explain at least part of the variation in the key response variable (AB_h). Analytically,

$$AB_h = \alpha + \beta IA_h + \gamma FC_h + \delta PC_h + \varepsilon_h \qquad (1)$$

where β, γ and δ are the parameter vectors to be estimated in order to assess respectively the role of individual farmers' attributes (IA_h), farm structural characteristics (FC_h), and farmers' psychological constructs (PC_h). $\varepsilon_h \sim N(0, \sigma^2)$ are the error terms, while α is the intercept. Figure 1 shows the frequency distribution of farmers' stated intention to crop giant reed, with a 0 hectare response expressed by about 8% of the respondents. Since null intention expressed by farmers assumes the typical corner-solution value of zero, the empirical model employed has to consider the presence of a censored outcome.

Ordinary Least Square (OLS) regression does not yield consistent parameter estimates due to a censored sample that is not representative of the population. In the Tobit model the regression is specified as an unobserved latent variable, AB^*_h:

$$AB^*_h = \alpha + \beta IA_h + \gamma FC_h + \delta PC_h + \varepsilon_h \qquad (2)$$

If AB^*_h were observed, we would estimate the parameters by OLS. The observed variable AB_h is related to the latent AB^*_h through the observation rule:

$$AB_h = \begin{cases} AB^*_h & \text{if } AB^*_h > L \\ L & \text{if } AB^*_h \leq L \end{cases} \qquad (3)$$

The probability of an observation being censored is

$$\begin{aligned} Pr(AB^*_h \leq L) &= Pr(\alpha + \beta IA_h + \gamma FC_h + \delta PC_h + \varepsilon_h \leq L) \\ &= \Phi\{(L - \beta IA_h + \gamma FC_h + \delta PC_h)/\sigma\} \end{aligned} \qquad (4)$$

where $\Phi(\cdot)$ is the standard normal cumulative distribution function.

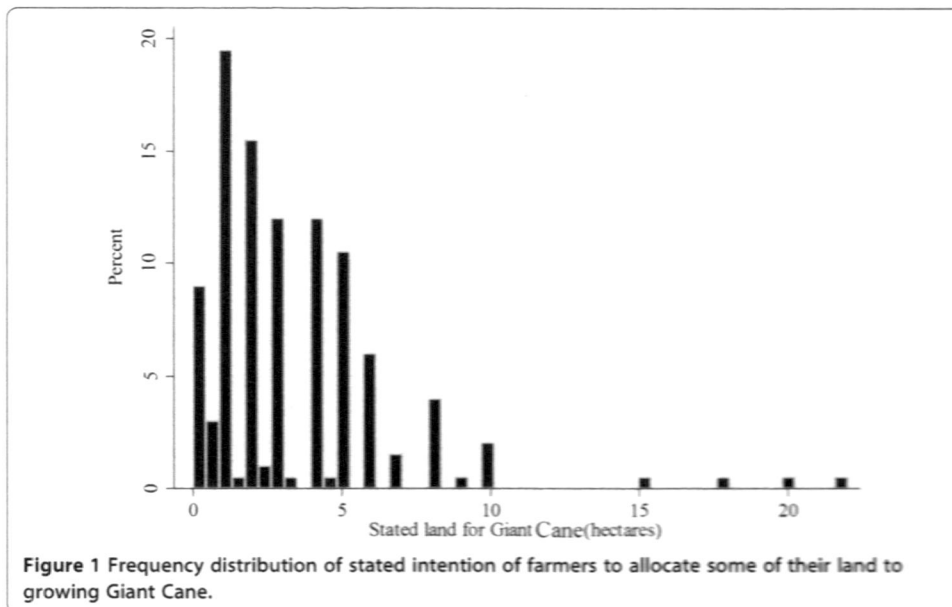

Figure 1 Frequency distribution of stated intention of farmers to allocate some of their land to growing Giant Cane.

Results and discussion

In an attempt to condense the set of information collected in the questionnaire section concerning the psychological constructs, a principal component analysis (PCA) was performed. PCA allows original data to be converted into latent constructs (or dimensions). Put differently, dimensions obtained by using PCA represent the main psychological constructs that serve to predict the pro-social (or cooperative) behavior of the sample of farmers interviewed.

The PCA results identified, based on the original 12 variables, five principal components explaining 69.3% of total variance (Table 4). The first construct (PC_1) was termed *social dilemma*. It is characterized by variables that detect a dilemma concerning environmental protection, or the ability to cooperate with other farmers facing environmental issues. The second extracted component (PC_2) was termed *awareness of responsibility*. This construct is positively characterized by variables related to personal credence which serve to provide a useful contribution to solutions to environmental issues that entail the production of renewable energy sources. The third construct (PC_3) was *personal norms*, based on variables underlining a moral obligation to perform or refrain from specific actions. The fourth component (PC_4), namely *trust in other people's cooperation*, comprises variables defining trust in other farmers' willingness to take part in collective action aimed at biomass production. This component is characterized, moreover, by a variable that identifies the trust that a farmer receives from his/her social network when deciding to contribute to environmental protection by producing agro-biomass. Last, but not least, construct (PC_5), or *awareness of consequences*, is identified by variables underlining the farmer's consciousness of the positive consequences that his/her pro-social behavior has, or may have, on other members in the area. Once calculated, the PCA scores of the psychological constructs (PC_1, PC_2, PC_3, PC_4, PC_5, PC_6) are included as predictors of the farmers' intentions to participate in the bioenergy supply chain (Table 5).

Table 6 reports the descriptive statistics of the independent variables selected in the study, while Table 7 shows estimates of the relationship between farmers' intentions to join the bioenergy supply chain and the main driving factors.

Table 5 Components matrix based on NAM variables

Variables	NAM - Psychological constructs	Components				
		PC_1	PC_2	PC_3	PC_4	PC_5
NAM3	Social dilemma	.840	-.058	.085	-.070	-.095
NAM2		.799	-.067	-.049	-.122	.062
NAM1		.729	.179	-.171	.074	-.039
NAM6	Awareness of responsibility	-.086	.843	.137	-.087	.076
NAM4		.212	.773	.037	.174	-.201
NAM5		-.082	.639	.097	.267	.418
NAM8	Personal/social norms	-.071	.111	.863	.012	.108
NAM7		-.038	.104	.708	.357	.076
NAM9	Trust in other people's cooperation	-.059	.019	.060	.885	.048
NAM10		-.060	.184	.388	.597	.094
NAM11	Awareness of consequences	.004	-.158	.384	-.195	.728
NAM12		-.053	.186	-.054	.372	.717

Table 6 Descriptive statistics of variables included in the empirical model

Variable	Description	Mean	Std.dev	Min	Max
AB	Agricultural land to bioenergy (ha)	3.37	3.22	0	22
IA_1	1 if male, 0 female	0.61	N.A	0	1
IA_2	Age (years)	43.67	12.18	18	80
IA_3	1 if farmer, 0 entrepreneur	0.83	N.A	0	1
IA_4	1 if full-time commitment, 0 part-time	0.75	N.A	0	1
IA_5	1 presence of recent investments, 0 otherwise	0.7	N.A	0	1
IA_6	1 if crops were changed recently, 0 otherwise	0.18	N.A	0	1
IA_7	1 if farmer knows cooperatives, 0 otherwise	0.59	N.A	0	1
IA_8	1 if farmer has participated in cooperatives, 0 otherwise	0.34	N.A	0	1
FC_1	1 presence of fallow land, 0 otherwise	0.34	N.A	0	1
FC_2	1 presence of livestock, 0 otherwise	0.23	N.A	0	1
FC_3	Total agricultural area	19.64	28.14	1.5	340
PC_1	Social dilemma	0	1	−1.44	2.66
PC_2	Awareness of responsibility	0	1	−3.92	2.75
PC_3	Personal norms	0	1	−3.37	2.24
PC_4	Trust in other people's cooperation	0	1	−3.74	1.8
PC_5	Awareness of consequences	0	1	−5.98	1.64

According to the parameter estimates, those related to psychological constructs (δ) and farm characteristics (γ) significantly affect farmers' propensity to participate in bioenergy supply more than parameters related to farmers' socio-demographic characteristics (β). Indeed, among farmers' characteristics, only farmer status (positively) and past knowledge on cooperatives (negatively) affect farmers' willingness to join the bioenergy supply chain. Indeed, two out of three farmers' structural characteristics appear to impact farmers' choice: farms with a larger land endowment are more likely to allocate land to cultivating Giant Reed, while the presence of livestock seems to reduce farmers' propensity to join collective action.

Looking at the role of psychological constructs, four of the five postulated dimensions were found to influence the likelihood of farmers joining collective action. To be precise, our results show that the likelihood of participating in the bioenergy chain increases when the *social dilemma* dimension decreases. This result confirms what was stated by Ostrom (2000). Specifically, higher suitability of collective action exists for farmers who are willing to cooperate to solve social and environmental dilemmas.

Estimates also indicate higher willingness to join collective action for farmers who consider themselves more useful in providing solutions to environmental issues by producing renewable energy sources (*awareness of responsibility* dimension). Finally, both farmers' moral obligation to specific pro-social engagement, or *personal norms*, and farmer's consciousness of the positive consequences of their pro-social behavior (*awareness of consequences*) seem to affect positively their willingness to participate in collective action.

Conclusions

The present study aimed to ascertain which determinants most affect farmers' willingness to contribute to common resources management. The empirical study was conducted in an area in the province of Avellino (southern Italy) highly affected by soil

Table 7 Estimates (in bold estimate of statistically significant parameters)

Parameter	Description	Estimates	t	P > t
β_1	Gender	0.136	0.980	0.330
β_2	Age	0.008	1.410	0.162
β_3	**Farmer or entrepreneur**	**0.867**	**4.040**	**0.000**
β_4	Full-time or part-time commitment	0.133	0.760	0.447
β_5	Presence of fallow land	0.128	0.840	0.404
β_6	Presence of livestock	−0.045	−0.260	0.799
β_7	**Presence of recent investments**	**0.504**	**3.250**	**0.001**
β_8	Crops were changed recently	−0.049	−0.300	0.764
γ_1	Farmer knows cooperatives	0.197	1.180	0.239
γ_2	**Farmer has participated in cooperatives**	**−0.289**	**−1.880**	**0.062**
γ_3	**Total agricultural area**	**0.010**	**4.350**	**0.000**
δ_1	**Social dilemma**	**−0.221**	**−3.090**	**0.002**
δ_2	**Awareness of responsibility**	**0.142**	**2.070**	**0.040**
δ_3	**Personal/social norms**	**0.114**	**1.660**	**0.099**
δ_4	Trust in other people's cooperation	−0.002	−0.030	0.973
δ_5	**Awareness of consequences**	**0.128**	**1.990**	**0.048**
α	**Intercept**	**−1.044**	**−3.050**	**0.003**

erosion problems. The study focused on the willingness of farmers to contribute to the public good through biomass production (giant reed). Overall estimates suggest that there is a systematic effect of the farmers' psychological constructs in driving farmers' behavior to be involved in the development of the agro-energy supply chain. The role of these dimensions is at least as important as farms' structural characteristics (land area and farming system). Less significant impact was obtained for farmers' individual characteristics.

The theoretical approach implemented in the paper appears confirmed by the case study results. Common resources management can be achieved when rules regarding the use of public goods are defined by involving stakeholders with strong interdependency (Ostrom, 2010). Social norms help to reduce free-riding by building reciprocity and trust. Moreover, social norms affect the behavior of individuals who are willing to join collective action, with, as a consequence, a sustainable use of public goods. This particularly holds when individuals, in the presence of a social dilemma, operate in the same context and share the same objectives. When such conditions take place, cooperation is initiated. In order to observe the long-lasting success of common resources management, others have to learn to trust the complex network of stakeholders and have to be more willing to adopt reciprocity leading to higher levels of cooperation. Hence it emerges that social dilemma and trust in other people's cooperation represent the most important determinants that affect long-lasting collective action. However, as a result of this study "trust in other people's cooperation" has no significant impact on willingness to participate in collective action to build a bioenergy supply chain. This could be due to the fact that our study concerns an early stage collective action in which interactions are not in place. Put differently, while the trust variable makes sense in the theoretical framework, in this study there are no interactions among actors to be tested.

Results do not have direct policy or agribusiness implications. However, it captures the existence of conditions able to develop a collective action, through cooperation, aiming at common pool resources management.

Starting from the concept introduced on the basis of this case study, future research could develop in-depth studies in at least two directions. First, the area where data were collected is limited to one region in southern Italy. We believe the area in question is representative of the many Mediterranean areas affected by both low income and environmental issues. However, replication of this study in other areas is necessary to test the external validity of our results. Second, this study tested the participation in collective action on an individual farmer's non-coordinated decision. The results could be different in the case of cooperation induced by external coordination. In this respect, some aspects of collective action, common norms in particular, would be assured by the cooperation and the results could be different. Finally, Ostrom (1990) describes collective action as a dynamic perspective. Future research should take this aspect into account.

Better knowledge of collective action and farmers' propensity to undertake such action could be useful to promote long-term sustainable management of public goods. Therefore the role of political institutions should be to accompany such processes in order to facilitate networking by creating agricultural policies that promote local bioenergy production and sustainable management of public goods with a view to reducing pollution emissions (Kotchen and Moore, 2007) and mitigating soil erosion risk. After all, the call for sustainable development needs to be based on new civic values as well as new forms of participation.

Competing interests
The authors declare that they have no competing interests.

Authors' contribution
Authors are equally responsible of every paragraph of the paper. All authors read and approved the final manuscript.

Acknowledgements
This research received grant from the European Regional Development Fund (PON): "Integrated agro-industrial chains with energy efficiency for the development of eco-compatible processes of energy and bio-chemical production for renewable sources for the land valorization (ENERBIOCHEM)".

Author details
[1]Department of Agriculture, AgEcon and Policy Group, University of Naples Federico II, Via Università 96, Portici, NA 80055, Italy. [2]Department of Agricultural and Forest Sciences, AgEcon and Policy Group, University of Palermo, Viale delle Scienze ed. 4, Palermo 90128, Italy.

References
Agrawal A, Goyal S (2001) Group size and collective action: third party monitoring in common pool resources. Comp Pol Stud 34(1):63–93
Antinori C, Bray DB (2005) Community forest enterprises as entrepreneurial firms: economic and institutional perspectives from Mexico. World Dev 33(9):1529–1543
Berkes F (1987) Common Property Resource Management and Cree Indian Fisheries in Subarctic Canada. In: McCay BJ, Acheson J (eds) The Question of the Commons: The Culture and Ecology of Common resources. University of Arizona Press, Tucson, pp 66–91
Blamey R (1998) Contingent valuation and the activation of environmental norms. Ecol Econ 24:47–72
Caracciolo F, Lombardi P (2012) A new-institutional framework to explore the trade-off between Agriculture, Environment and Landscape. Econ Policy Energy Environ 3:135–154
Dawes RM, Orbell JM, Simmons RT, Van De Kragt AJC (1986) Organizing groups for collective action. Am Polit Sci Rev 80(4):1171–1185
De Groot JI, Steg L (2009) Morality and prosocial behavior: The role of awareness, responsibility, and norms in the Norm Activation Model. J Soc Psychol 149(4):425–449

Ebreo A, Vining J, Cristancho S (2003) Responsibility for environmental problems and the consequences of waste reduction: A test of the norm-activation model. J Environ Syst 29(3):219–244

Fujiie M, Hayami Y, Kikuchi M (2005) The conditions of collective action for local commons management: the case of irrigation in the Philippines. Agric Econ 33(2):179–189

Gibson CC, Williams JT, Ostrom E (2005) Local enforcement and better forests. World Dev 33(2):273–284

Guagnano GA (2001) Altruism and market-like behavior: an analysis of willingness to pay for recycled paper products. Popul Environ 22(4):425–438

Guagnano GA, Dietz T, Stern PC (1994) Willingness to pay for public goods: a test of the contribution model. Psychol Sci 5(6):411–415

Guagnano GA, Stern PC, Dietz T (1995) Influences on attitude-behavior relationships: a natural experiment with curbside recycling. Environ Behav 27:699–718

Hisschemöller M, Sioziou I (2013) Boundary organizations for resource mobilization: enhancing citizens' involvement in the Dutch energy transition. Environ Politics 22(5):792–810

Hopper JR, Nielsen JM (1991) Recycling as altruistic behavior. Normative and behavioral strategies to expand participation in a community recycling program. Environ Behav 23:195–220

Joireman JA, Lasane TP, Bennett J, Richards D, Solaimani S (2001) Integrating social value orientation and the consideration of future consequences within the extended norm activation model of proenvironmental behaviour. Br J Soc Psychol 40(1):133–155

Kotchen MJ, Moore MR (2007) Private provision of environmental public goods: Household participation in green-electricity programs. J Environ Econ Manag 53(1):1–16

Liebe U, Preisendörfer P, Meyerhoff J (2010) To pay or not to pay: Competing theories to explain individuals' willingness to pay for public environmental goods. Environ Behav 43(1):106–130

Liebe U, Preisendörfer P, Meyerhoff J (2011) To Pay or Not to Pay: competing theories to explain Individuals' willingness to Pay for public environmental goods. Environ Behav 43(1):106–130

Meinzen-Dick R, Raju KV, Gulati A (2002) What affects organization and collective action for managing resources? Evidence from canal irrigation systems in India. World Dev 30(4):649–666

Migliore G, Schifani G, Dara Guccione G, Cembalo L (2014) Food Community Networks as Leverage for Social Embeddedness. J Agric Environ Ethics 27(4):549–567, DOI: 10.1007/s10806-013-9476-5, 1–19

Migliore G, Schifani G, Cembalo L (2015) Opening the black box of food quality in the short supply chain: Effects of conventions of quality on consumer choice. Food Qual Prefer 39:141–146

Montada L, Kals E (2000) Political implications of psychological research on ecological justice and proenvironmental behaviour. Int J Psychol 35:168–176

Olson M (1965) The Logic of Collective Action: Public Goods and Theory of Groups. Harvard University Press, Cambridge MA

Ostrom E (1990) Governing the Commons: The Evolution of Institutions for Collective Action. Cambridge University Press, Cambridge, UK

Ostrom E (1991) Governing the Commons. The Evolution of Institutions for Collective Action. Cambridge University Press, Cambridge, MA

Ostrom E (1999) Social capital: a fad or a fundamental concept? In: Dasgupta P, Serageldin I (eds) Social capital: A multifaceted perspective. The World Bank, Washington, DC, pp 172–214

Ostrom E (2000) Collective action and the evolution of social norms. J Econ Perspect 14(3):137–158

Ostrom E (2010) Revising theory in light of experimental findings. J Econ Behav Organ 73:68–72

Ostrom E, Walker J (2003) Trust and Reciprocity: Interdisciplinary Lessons for Experimental Research, Volume VI in the Russell Sage Foundation Series on Trust, Russell Sage Foundation

Pretty J, Ward H (2001) Social capital and the environment. World Dev 29(2):209–227

Rootes C (2012) Climate change, environmental activism and community action in Britain. Soc Altern (Special issue on Community Climate Action) 31(1):24–28

Sayfang G (2010) Community action for sustainable housing: Building a low-carbon future. Energy Policy 38:7624–7633

Schwartz SH (1977) Normative influences on altruism. Adv Exp Soc Psychol 10:221–279

Schwartz SH, Howard JA (1980) Explanations of the moderating effect of responsibility denial on the personal norm-behavior relationship. Soc Psychol Q 43:441–446

Schwartz SH, Howard JA (1981) A normative decision-making model of altruism. Altruism and helping behavior, pp 189–211

Stern PC, Dietz T, Abel T, Guagnano GA, Kalof L (1999) A value-belief-norm theory of support for social movements: The case of environmentalism. Human Ecol Rev 6:81–95

Vining J, Ebreo A (1992) Predicting recycling behavior from global and specific environmental attitudes and changes in recycling opportunities. J Appl Soc Psychol 22:1580–1607

6

Crop substitution behavior among food crop farmers in Ghana: an efficient adaptation to climate change or costly stagnation in traditional agricultural production system?

Zakaria A Issahaku[1*] and Keshav L Maharjan[2]

* Correspondence:
zakh2000us@gmail.com
[1]Debt Management Division,
Ministry of Finance, P.O. Box MB 40
Accra, Ghana
Full list of author information is
available at the end of the article

Abstract

This study analyzes impact of climate change on yield, planting decisions and output of five major food crops (cassava, maize, sorghum, rice and yam) in Ghana. Results of Multivariate Tobit Model show that yield, planting decisions and output of cassava, maize, sorghum and rice will increase as a result of climate change. This is in clear contrast to the hypothesis that warming and drying will reduce crop yields in countries located within the tropics. Climate change impact on yields, planting decisions and output of yam is projected to be negative. Analysis of planting decisions indicates that climate change will stimulate farmers to allocate more land for cassava, maize, sorghum and rice cultivation. It is observed that farmers respond to positive impact of climate on yields of cassava, maize, sorghum and rice by reallocating more land for the cultivation of these crops, which is in line with neoclassical understanding of producer behavior. In contrast, farmers' response to price rise does not display a consistent pattern. By and large, farmers respond weakly to price increases. This peculiar trait of food crop farmers can stifle the future development of the food crop subsector in Ghana.

Keywords: Multivariate Tobit Model; Climate change; Major food crops; Crop yield; Farm size; Ghana

Background

Climate is one of the most important inputs in agricultural production system. The hard truth that the world is getting hotter and drier is of a major concern to many a developing country dependent on agriculture as their main livelihood source. Low precipitation and increased frequency of extreme events such as floods and droughts can reduce crop yields and increase risks in agricultural production in many countries located in the lower latitudes. In Ghana, agriculture is largely rain-fed, and the vagaries of the weather determine agricultural productivity. Farmers usually respond to reduction in crop yield by putting more land into cultivation. It is therefore no wonder that yield levels of major food crops are significantly lower than their potential levels, indicating a potential of raising outputs of major food crops through crop productivity growth. Cassava, maize, sorghum, rice and yam have yield gaps of 57.5%, 40%, 33.33%, 40% and

38%, respectively (MOFA 2007). Increasing agricultural growth by land expansion may not be sustainable because farmers are not only limited by plot size in their possession but also difficulties associated with managing large tracts of land under cultivation including labor availability and loss of forest cover. Increasing production of major staple crops can be enhanced by utilizing the land more intensively thereby closing these crop yield gaps (Breisinger et al. 2009).

Crop intensification is an option mainly for commercial farmers, since they are more likely to be linked to national markets and international agribusinesses and be able to invest in agricultural technologies. Smallholders in developing countries do not have much access to agricultural inputs such as fertilizers, pesticides and improved seeds. Besides crop intensification, agriculture production growth could also be reached using sustainable agriculture technologies which are extensively promoted by international development agencies and research centers (Branca et al. 2013; Garrity et al. 2010). It is imperative to note that adopting intensive farming and other farm practices is not free from the adverse effects of changing climate. More recent literature points to the adverse impact of changing climate on crop productivity. A review of climate impact literature on various crops by Knox et al. (2012) indicates that yields of cassava, sorghum, millet and maize will decrease in West Africa through adverse effects of climate change. Warming and drying exacerbate stresses in crop plants, potentially leading to catastrophic yield reductions: It reduces water availability for irrigation; it also reduces soil fertility through increased oxidation of soil organic carbon; and it also increases incidence of pests, diseases and weeds. Sagoe (2006) used crop simulation model to analyze climate change impact on root crops in Ghana and the results indicate reductions in yields of cassava and cocoyam under all projected climate scenarios. Analysis of projected climate change impact in Ghana's initial communication to Inter-governmental Panel on Climate Change (IPCC) also indicates reductions in yields of maize in the transition zone, located between the forest and the savanna ecological zones in Ghana (GEPA 2001). The afore-mentioned analyses and other similar studies are based on crop simulation models which show relationship between environmental variables including climate and the growth of crop plants. The effect of climate on crop yield may be more complex than just mere climate-crop plant growth relationship. Other factors can reverse an otherwise positive or negative effect of climate on crop yield. The failure to take into account the role of non-environmental variables denoting farm or farmer characteristics and/or management practices by farmers may undermine the use of crop simulation models in climate research. Further, research on the impact of climate change on agricultural production has mainly focused on the effect of climate and its variability on individual crops, while the potential for adapting to climate change through crop substitution has received less attention. Crop switching has not generally been captured by the climate change.

Based on national survey data, this paper intends to extend this line of analysis by using the production function approach, which considers and incorporates some socioeconomic variables in analyzing climate impact not only on crop yield, but also on farmers' planting decisions. Analysis of this nature is intuitive because food farmers especially in developing countries may not respond to price/profit incentive but rather higher yield or output. This therefore makes necessary it to investigate the alignment of yield-maximizing behavior of food crop farmers with planting decisions. In the next section, a review of some adaptive responses to adverse on-farm conditions by farmers is carried out. Econometric methods used to assess the effect of various factors including climate variables on crop production are explained in detail in section Methods using appropriate mathematical equations. In addition, the data and summary statistics of all variables used in this study are also presented and explained. In section Results and discussion, model results are presented and discussed. In section Climate change impact on food crop production in Ghana, regression coefficients together with trend of climate variables are used to simulate the impact of changing climate on crop yield and planted area in future years. The last section presents the discussion and summary of the research findings and proceeds to make recommendations for consideration of policymakers.

Review of farm-level adaptation measures

Climate change is predicted to have negative impact on farm outcomes in most developing countries. The potential yield- or welfare-reducing effect of climate change can be ameliorated by autonomous adaptation by farmers. Climate change impact studies which ignore farmers' adaptive responses are likely to overstate damages or understate benefits of climate change (Mendelsohn et al. 1994).

Broadly speaking, farmers adapt to varying climate either through the modification of the set of crops they choose to plant or improved cultural/management operations. Although many studies assume that no change in future in the set of crops cultivated, the importance of adopting crops or varieties which are tolerant to projected warmer and drier climate in the future cannot be glossed over. Adaptation options accessible by farmers include crop diversification, varying crop varieties, changing planting and harvesting dates, irrigation, use of water and soil conservation techniques and diversifying away from farming (Nhemachena and Hassan 2007). A study by Corobov (2002) in Moldova shows that the adoption of late-maturing maize hybrids as an adaptation measure engenders considerable yield enhancement. The adoption of heat tolerant varieties of sorghum, millet, cotton, cowpeas and rice in Mali also attest to the yield- and welfare-enhancing effects of this adaptation measure (Butt et al. 2005). Additionally, farmers can response to climate change in a more radical manner by changing the mix of crops grown. Kurukulasuriya and Mendelsohn (2006) and Seo and Mendelsohn (2008a) climate change adaptation studies for Africa and South America, respectively, have demonstrated how farmers change their planting decisions in response to warmer and drier climate. In dry and warm locations of Africa, farmers tend to choose millet and groundnut while maize and beans are chosen in wet but warm locations. Similarly, in warm locations of South America, farmers grow squash, fruits and vegetables while potatoes and maize are grown in cooler locations. Varying set of crops cultivated induces spatial shift of production pattern to places with mild climate. An assessment of global impact of climate change by Darwin et al. (1995) indicates that expansion of cultivable land will increase global food production. Butt et al. (2005) reports that the shift in crop production patterns southwards in Mali has helped to mitigate welfare losses.

The use of crop switching as adaptation measure is hindered by the fact that production decisions by farmers may not be motivated by yield- or profit-optimizing behavior. Farmers as consumers may have certain preferences for some crops over others. They are therefore likely to choose these crops although they may not be the ideal crop in terms of yield or welfare maximization (Chipanshi et al. 2003). The disposition of farmers to risk may also stifle crop-switching efforts. Small-scale farmers are usually risk-averse and they are more likely to settle for crops with low yield variance rather than riskier but high yielding crops (Kaiser et al. 1993). Finally, spatial shift in cropping patterns may result in degradation of tropical rainforest with its attendant effects on the ecosystem and loss of some economic benefits from utilization of forest resources (Darwin et al. 1995).

Methods

Empirical model

This study is based on the notion that climate is one of the important determinants of crop productivity. Most climate impact studies tend to focus on crop yield with the assumption that farmers hardly vary their cropping decisions, implying that a crop yield change proportionately translates to crop output change. Since crop output is a product of crop yield and harvested or planted area, this study evaluates the impact of climate on both yield and planted area of major food crops in Ghana in order that the combined effect on crop output can be determined.

The first step in assessing potential costs and climate change adaptation strategies is to determine the effect of climate on crop yields (Cabas et al. 2010). One of the methods to measure the sensitivity of crop yields to changing climate is to analyze how actual crop yields vary across different locations with different climatic conditions (Mendelsohn and Dinar 2009). Regression models have the potential flexibility to integrate both physiological determinants of yield including climate and socioeconomic factors. With this approach, an appropriate production function is specified in order to isolate the effect of climate from the effects of other confounding variables including modern inputs and socioeconomic variables. Formally, a production function developed by Just and Pope (1978) to analyze effect of production inputs on crop yields is specified as in equation (1).

$$Y^* = f(X, \beta) + \mu \tag{1}$$

$$Y = \begin{cases} Y^* \text{ if } Y^* > 0 \\ 0 \text{ if } Y^* \leq 0 \end{cases} \tag{2}$$

Y^* is a latent variable for observed crop yield, Y; X is vector of independent variables; μ is stochastic error term for crop yield model which is assumed to be multivariate normally distributed. The symbol β represents vector of parameters for the model.

Equation (1) can be estimated for each crop individually or jointly for all crops in question. In Ghana, farmers usually cultivate more than more than one crop on a plot of land. It is therefore difficult to isolate the effect of production inputs on each crop as crops are produced jointly. In this instance, a system of single production functions can be estimated jointly for all mixed crops. Estimating production functions jointly is appropriate when the vector of explanatory variables is same across crops, and cross-equation correlations are relevant (Zellner 1962). As a result, the production functions in this study are estimated jointly for cassava, maize, sorghum, rice and yam at the farm level using Multivariate Tobit (MVT) specification as in equations (1).

The second strand of analysis of the impact of climate variables on crop production is its impact on planted area. Amount of land allocated to each crop has a substantial influence on the level of production. Acreage of planted area allocated will only be positive for those crops an individual farmer decides to plant. That is, there is a selection bias as a result of the decision by a reasonable number of farmers not to cultivate some of the crops considered under this study, otherwise known as corner solution in econometric parlance. More formally, this type of data can be fitted using Tobit model specification as in equations (3) and (4).

$$A^* = f(Z, \gamma) + \varepsilon \tag{3}$$

A^* is the latent planted farm area; Z is the vector of explanatory variables; γ is the vector of regression coefficients; and ε is the error term for the planted area allocation model with multivariate normal distribution. The observed planted farm area for crops (A) is related to the latent planted area as in equation (4).

$$A = \begin{cases} A^* & \text{if } A^* > 0 \\ 0 & \text{if } A^* \leq 0 \end{cases} \tag{4}$$

Equation (3) is estimated for the five major food crops in question simultaneously with the notion that unobserved factors that influence crop planted area allocation are correlated. That is, a MVT model also is employed as it provides efficient statistical estimation of parameter estimates under the conditions stated above. Since there are five crop choices involved, it will therefore mean that evaluation of the five-dimensional multivariate normal integral will be a difficult exercise. To improve estimation time and accuracy, the maximum simulated likelihood (MSL) is employed using Geweke-Hajivassiliou-Keane (GHK) simulator in estimating the model in equations (1) and (3) (Roodman 2011; Hajivassiliou et al. 1996).

The impact of future climate on crop yield, planted area and crop output is simulated using equations (5), (6) and (7), respectively.

$$\Delta Y\% = \frac{(Y_{future} - Y_{present})}{Y_{present}} \times 100 \tag{5}$$

$$\Delta A\% = \frac{(A_{future} - A_{present})}{A_{present}} \times 100 \tag{6}$$

$$\Delta Q\% = \Delta Y\% + \Delta A\% \tag{7}$$

$\Delta Y\%$ is percentage change in crop yield; Y_{futute} is the future prediction of crop yield; $Y_{present}$ is the predicted crop yield in the current period; $\Delta A\%$ is change in planted area; A_{futute} is the future prediction of planted area; $A_{present}$ is the predicted planted area in the current period; and $\Delta Q\%$ is the percentage change in crop output.

Data and descriptive statistics

This study analyzes the effects of climate variables on crop production using data from fifth round of Ghana Living Standards Survey (GLSS V) compiled by Ghana Statistical Service (GSS) in 2005/2006 as well as climate (temperature and rainfall) data sourced from Ghana Meteorological Service Agency for ten weather stations across the length and breadth of the country. The GLSS V data contains information on socioeconomic characteristics of 8,687 households. For the purpose of this study, 2,577 farming households which cultivate at least one of the five major crops of cassava, maize, sorghum, rice and yam are considered.

Five different crop yields expressed in kg/ha are used as the dependent variables: cassava, maize, sorghum, rice and yam. Crop yield is calculated by dividing total crop output by hectares of harvested farm area. Yields of the major food crops are generally low. The mean crop yields range from a low of about 346.33 kg/ha for sorghum to a high of about 5,342.36 kg/ha for cassava as shown in Table 1. Additionally, farm sizes for the five crops are also used as dependent variables. The mean farm size ranges from 1.65 hectares for cassava to 3.22 hectares for sorghum. The independent variables used in the study include climate variables (minimum and maximum temperature, and rainfall), household labor, age, gender and education of household head, crop prices and farm input (Table 1).

Table 1 Description and summary statistics of model variables

Variables	N	Mean	Standard deviation	Minimum	Maximum
Crop yield (kg/ha)					
Cassava	1513	5342.36	8437.83	45.59	85097.14
Maize	2017	661.26	1429.954	0.59	32592.59
Sorghum	560	346.33	503.37	2.70	4325.26
Rice	333	637.65	1154.79	2.06	12722.65
Yam	682	3150.98	5935.54	61.31	72598.55
Farm size (ha)					
Cassava	1508	1.65	3.27	0.02	97.10
Maize	2012	1.97	4.07	0.02	97.10
Sorghum	555	3.22	10.69	0.09	194.21
Rice	328	2.13	4.07	0.05	42.48
Yam	677	2.01	4.14	0.04	82.94
Output price (GHS/kg)					
Cassava	2577	0.12	0.07	0.02	1.07
Maize	2577	0.55	0.84	0.01	6.00
Sorghum	2577	0.38	0.40	0.12	4.00
Rice	2577	0.41	0.59	0.03	9.00
Yam	2577	0.46	0.32	0.07	5.47
Climate					
Minimum temperature (°C)	2577	22.29	0.63	21.18	23.48
Maximum temperature (°C)	2577	31.88	1.66	29.6	34.94
Rainfall (mm)	2577	1179.79	145.95	807.69	1377.49
Socioeconomic					
Household labor (number)	2577	2.58	1.64	0	16
Age-head (years)	2577	45.57	14.62	18	95
Gender-head (=1 if female)	2577	0.17	0.38	0	1
Education-head (years)	2577	3.13	4.60	0	16
Farm input (GHS/ha)	2577	35.79	84.81	0	1200

Notes GHS: Ghana cedi; 1 US$ = 0.92 GHS as of December 2005.
Source: calculated from 2005 Ghana Living Standard Survey and Ghana Meteorological Agency data.

The climate variables used in this study are normal minimum and maximum temperature and normal annual rainfall for the crops in question. The climate data covers fifty years (1961-2010), a long enough period to be used to construct normal climate variables. The climate data is, in turn, matched with locations of farming households as identified in the GLSS V. Ghana is generally a warm country with high temperature all year round. Normal temperature during the effective growing season of crops is about 26 °C. It is hypothesized that high temperature will impact negatively on yields of the crops in question. Normal monthly rainfall during effective growing season for crops is about 1179.79 mm per month. Since all crops need wet conditions up to a certain threshold, it is hypothesized that rainfall will have positive effect on all crops. The non-climate variables including household labor, age and education of household head and farm inputs have hypothesized positive effect on crop yield while gender has hypothesized negative effect on crop yield. The average household labor force, defined as the number of household members aged 15-60, is about 3 indicating the contribution of family labor to on-farm production. Average age of family heads is about 46 years who are mostly males with about three years of formal education. Although majority of Ghanaian food crop farmers cultivate cassava and maize, they do so on a plot size of about 2 hectares.

Farm inputs, which is the sum of household expenditure of hired labor, fertilizer, pesticide, seeds, irrigation and machinery are also hypothesized to have positive effect on all crop yields since their enhanced use are likely to increase crop yield. Generally speaking, farmers do not adequately use farm inputs in Ghana for various reasons including weak financial position. The average expenditure on farm inputs is about GHS 35.79. The meager expenditure on farm inputs reflects the fact that a significant number of farmers do not purchase these input at all.

Results and discussion

Climate effects on crop yield

This section presents and discusses the results of the MVT model from equation (1). Before the estimation was carried out, some validity checks of model data were undertaken to ensure that no outliers skew expected model results. Diagnostic tests indicate that model has no multicollinearity problems. Breusch-Pagan/Cook-Weisberg test, however, revealed presence of heteroskedasticity, implying that analyzing this data without addressing this problem would result inefficient parameter estimates, although the estimates would still be unbiased. This model is therefore estimated using as a robust regression, which helps to resolve the problem of heteroskedasticity. Ramsey RESET test for omitted variables bias in the MVT regression reveals no problems of misspecification for all regressions. The results of running MVT regression for crop yields on the set of the independent variables selected for this study as in equation (1) is displayed in Table 2. Wald chi^2 is statistically significant at 1% indicating the independent variables used in this model jointly provide plausible explanation for crop yield trends.

Household labor has statistically significant positive effect on yields of sorghum and yam. Its effects on yields of maize and rice have correct hypothesized signs but they are not statistically significant. Family members are engaged in on-farm activities including clearing, sowing, weeding and harvesting of these crops during the growing season when demand for alternative labor sources are high in farming communities. Age of household

Table 2 Results of crop yield Multivariate Tobit regression

Variables	Cassava	Maize	Sorghum	Rice	Yam
Intercept	−22085.7*** (6979.3)	−1164.1** (1294.1)	−326.04*** (274.58)	−1627.0*** (466.52)	6394.3 (3353.6)
Minimum temperature (°C)	1343.5*** (161.98)	37.157 (23.640)	−33.917*** (8.2782)	−14.726 (20.900)	−598.40*** (144.27)
Maximum temperature (°C)	−422.27*** (76.506)	9.4009 (10.171)	44.748*** (4.8346)	65.896*** (9.8528)	255.50*** (47.226)
Rainfall (mm)	6.7402*** (0.8432)	0.4223*** (0.1081)	−0.2217*** (0.0387)	−0.0365 (0.0932)	−0.6704 (0.4607)
Household labor	−0.3968 (63.653)	9.2969 (6.2705)	3.6002** (1.7373)	4.2471 (5.2064)	84.960*** (26.098)
Age of household head	12.709** (6.4873)	−0.4100 (1.4013)	−0.0065 (0.1929)	−0.5743 (0.6322)	6.8758** (3.3654)
Female household head	−47.711 (213.04)	−21.759 (35.012)	−19.482* (11.810)	−82.135** (31.884)	−370.89*** (140.82)
Education of household head	2.6794 (20.073)	2.0806 (3.5869)	−1.7510* (0.9272)	−1.1015 (2.4904)	−15.585 (10.595)
Farm input	5.6789** (2.6433)	1.3230*** (0.4452)	−0.2184*** (0.0834)	0.3579*** (0.1118)	0.1537 (0.5131)

Notes: *** means significant at 1%, ** means significant at 5% and * means significant at 10%; the total number of observations = 2577; Wald chi^2 (8) =111.34***; Log pseudo-likelihood =-31604170; the dependent variable is crop yield; and Figures in parenthesis are standard errors of regression estimates.

head has statistically significant positive effect on yields of cassava and yam. This means that older farmers who are engaged in cassava and yam cultivation achieve higher yields as compared to younger farmers. The coefficient of gender is negative and significant in the sorghum, rice and yam regressions but not statistically significant for other crops, implying that male headed homes gain higher yields from the cultivation of these crops but there is no statistical significant difference in yields of other food crop between male and female farmers. By and large, education of the household head has no significant effect on yields of all food crops. Researches in experimental plant physiology show that the soil fertility can be enhanced by adding several soil nutrient supplements to the soil (Ramteke and Shirgave 2012). Farm input has significant positive effect on yields of cassava, maize, sorghum and rice but its effect on yield of yam is not statistically significant.

Climate variables tend to have mixed effects on yields of food crops. Minimum temperature has significant positive on cassava yield but statistically negative effect on yields of sorghum and yam. Maximum temperature has significant negative effect on yields of cassava but positive effect on yields of sorghum, rice and yam. Both minimum and maximum temperatures have no statistically significant effect on yield of maize. Rainfall has significant positive effect on yields of cassava and maize but negative effect on sorghum yield. Positive effect of additional rains on yields of cassava and maize confirms the fact that cassava and maize are rain-loving crops which benefit from reasonably wet climatic conditions. Sorghum requires a reasonable amount of water from germination up till heading. Additional rains after heading can, however, be harmful. Although rainfall has no significant impact on rice yield, it may be true in rice growing areas in Ghana. Rice is mostly cultivated in the valley bottoms of the drier savanna ecological zone with waterlogged soils. As a result, additional rains may not bring about significant increase in yield of rice.

Impact of climate variables on planting decisions

This section analyzes the impact of climate variables on planting decisions of food crop farmers in Ghana using a Multivariate Tobit model, with farm size of each crop being used as dependent variable and, climate variables and crop prices as independent variables. These independent variables fit the model well as evidenced by significant values of Wald chi^2 (Table 3).

Prices of food crops do not display any consistent pattern in terms of its effect on land allocation (Table 3). Farmers who cultivate sorghum and rice respond positively to increases in the prices of these crops by putting more land into cultivation. Maize farmers, however, respond negatively to increase in the price of maize. Crop prices especially for cassava, maize and sorghum prices, however, have significant indirect effect on planting decisions. Increase in cassava price prompts farmers to allocate more land for maize and less land for sorghum and yam cultivation. Increase in maize price is associated with increased land allocated for cassava. . While sorghum price signal farmers to increase planted area allocated for yam cultivation, it, however, decreases the amount of land put into rice cultivation. Rice and cassava prices do not have significant cross-price effects.

Climate variables have statistically significant influence on farmers' planting decision. Minimum temperature has significant positive coefficients for cassava and significant

Table 3 Results of farm size Multivariate Tobit regressions

Variables	Cassava	Maize	Sorghum	Rice	Yam
Intercept	-3.6388** (2.6345)	-1.635921 (3.686328)	-7.7645** (5.1367)	-2.9370** (1.6151)	2.058012 (1.6034)
Minimum temperature (°C)	0.4814*** (0.0934)	0.0620 (0.0736)	-0.3945** (0.1789)	-0.2093*** (0.0734)	-0.2440*** (0.0764)
Maximum temperature (°C)	-0.2900 (0.0344)	0.0479* (0.0265)	0.6878*** (0.2135)	0.2567*** (0.0421)	0.1332*** (0.0361)
Rainfall (cm)	0.0026*** (0.0005)	0.0003 (0.0003)	-0.0032*** (0.0011)	-0.0003 (0.0003)	-0.0002 (0.0003)
Cassava price	0.1497 (0.1701)	0.4618* (0.2478)	-4.7550* (2.6651)	0.5283 (0.3677)	-1.1822* (0.6329)
Maize price	0.0607*** (0.0191)	-0.0681*** (0.0216)	-0.7577 (0.3776)	-0.0871** (0.0403)	-0.0578 (0.0426)
Sorghum price	0.0143 (0.0888)	-0.0868 (0.0788)	0.8904** (0.4143)	-0.1247* (0.0687)	0.2969*** (0.0641)
Rice price	0.0516 (0.0456)	-0.0479 (0.0381)	-0.3624 (0.3054)	0.1058*** (0.0357)	-0.0329 (0.0501)
Yam price	-0.0763 (0.0813)	-0.0834 (0.1046)	-0.3765 (0.3321)	0.1140 (0.0715)	-0.1982 (0.2084)

Notes: *** means significant at 1%, ** means significant at 5% and * means significant at 10%; the total number of observations = 2572; Wald chi^2 (8) = 97.98***, Log pseudo-likelihood = -10301403; the dependent variable is farm size; and Figures in parenthesis are standard errors of regression estimates.

negative coefficients for sorghum, rice and yam. This means that an increase in minimum temperature provokes reallocation of land away from sorghum, rice and yam towards cassava. Similarly, maximum temperature has significant positive coefficients for maize, sorghum, rice and yam. This means that an increase in maximum temperature increases land allocated for these crops. Rainfall has significant positive coefficients for cassava and negative for sorghum. This implies that reduced levels of rainfall will impact negatively on land allocated for the cultivation of sorghum while cassava benefits from additional rains through increased land allocation.

Climate change impact on food crop production in Ghana

This section simulates future change in climate using historical trend of climate variables (minimum and maximum temperature, and Rainfall) in order to analyze the impact of climate change on food crop production in Ghana. Trend analysis of climate variables over the period 1961-2010 show that both minimum and maximum temperatures are projected to increase while rainfall is projected to decline. Table 4 presents the future changes in temperature and rainfall generated using climate trends coefficients across the ten regions of Ghana. With the exception of Brong Ahafo region, rainfall is projected to reduce in all regions. Both minimum and maximum temperatures are projected to increase in all regions.

Using the change in climate variables as reported in Table 4 together with coefficients from the estimation of yield and farm size equations (Tables 2 and 3), we can simulate the future impact of climate change on food crop production.

It can be seen from Table 5 that climate change will raise the yields of cassava, maize, sorghum and rice but it will impact negatively on yields of yam. Cassava yield is expected to increase by 1.25%, 3.02% and 3.76% for 2015, 2020 and 2025, respectively. Maize yield is projected to go up by 0.33%, 0.86% and 0.99% for 2015, 2020 and 2025, respectively. Similarly, sorghum yield is projected to increase by 5.32%, 9.75% and 15.63% for 2015, 2020 and 2025, respectively. The yields of rice will increase by 8.79%, 17.50% and 26.30% 2015, 2020 and 2025, respectively. The yield of rice, however, will reduce by 3.57%, 7.32% and 10.74% for 2015, 2020 and 2025, respectively.

Table 4 Change in climate variables vis-à-vis 2010 values

	Rainfall (mm)			Minimum temperature (°C)			Maximum temperature (°C)		
	2015	2020	2025	2015	2020	2025	2015	2020	2025
Upper west region	−3.249	−6.498	−9.747	0.1095	0.219	0.329	0.102	0.203	0.305
Uppereast region	−0.583	−1.166	−1.749	0.1175	0.235	0.353	0.124	0.247	0.371
Northern region	−4.196	−8.391	−12.587	0.1065	0.213	0.320	0.136	0.271	0.407
Brong-Ahafo region	7.920	15.84	23.76	0.0615	0.123	0.185	0.126	0.251	0.377
Volta region	−16.269	−32.538	−48.807	0.1595	0.319	0.479	0.148	0.295	0.443
Ashanti region	−20.717	−41.433	−62.150	0.137	0.274	0.411	0.181	0.367	0.542
Eastern region	−12.887	−25.774	−38.661	0.101	0.202	0.303	0.148	0.295	0.443
Greater Accra region	−20.183	−40.365	−60.548	0.2165	0.433	0.650	0.095	0.189	0.284
Central region	−26.072	−52.143	−78.215	0.104	0.208	0.312	0.033	0.066	0.099
Western region	5.794	10.429	17.382	0.137	0.274	0.411	0.130	0.260	0.390

Table 5 Impact of climate change on food crop production (in percentage point change)

	2015	2020	2025
Crop yield			
Cassava	1.247	3.023	3.764
Maize	0.329	0.860	0.987
Sorghum	5.321	9.748	15.631
Rice	8.791	17.495	26.303
Yam	−3.565	−7.320	−10.739
Farm size			
Cassava	1.840	3.672	5.559
Maize	0.702	1.451	2.103
Sorghum	10.432	19.777	31.199
Rice	4.477	8.697	11.254
Yam	−1.781	−3.623	−5.378
Crop output			
Cassava	3.087	6.695	9.323
Maize	1.031	2.311	3.091
Sorghum	15.718	29.625	46.997
Rice	13.274	26.202	37.567
Yam	−5.351	−10.956	−16.132

Notes: all figures represent changes from the base year of 2010.

Climate change will prompt farmers to increase the cultivation of cassava, maize, sorghum and rice, as evidenced by the increased levels of land allocated for cassava, maize, sorghum and rice (Table 5). By 2015, climate change is projected to raise farm size for cassava, maize, sorghum and rice by 1.84%, 0.70%, 10.43% and 4.48%, respectively while farm size for yam will decrease by 1.78%. By 2025, land allocated for cassava, maize, sorghum and rice cultivation will increase by 5.56%, 2.10%, 31.20% and 11.25%, respectively, while decreasing farm size for yam by 5.38%.

Overall, climate change will increase output of cassava, maize, sorghum and rice by 3.09%, 1.03%, 15.72% and 13.27% by 2015. By 2020, climate change is projected to raise output of cassava, maize, sorghum and rice by 6.70%, 2.31%, 29.63% and 26.20%, respectively. By 2025, output of cassava, maize, sorghum and rice cultivation will increase by 9.32%, 3.09%, 47.00% and 37.57%, respectively %. Output of yam will, however, decline by 5.35%, 10.96% and 16.13% for 2015, 2020 and 2025, respectively.

Conclusions and discussion

This study uses national household survey data to analyze impact of climate change on food crop production in Ghana. Multivariate Tobit Model is used to assess the impact of climate variables on crop yields and planting decisions of farmers. It is found that climate change will have positive impact on yields, farm size and output of cassava, maize, sorghum and rice. This is in clear contrast to the hypothesis that climate change will reduce crop yields in countries located within the tropics. Yam, however, proved to be more susceptible to climate change. The use of farm inputs significantly improves yields of food crops. But, the effects of these inputs are economically smaller as compared to climate variables. Apart from climate variables and farm inputs, some socioeconomic variables, especially household labor and gender of household heads also have significant influence on crop yields.

Further, climate change has significant influence on crop planting decisions. Additional warming and drying will prompt farmers to increase the cultivation of cassava, maize, sorghum and rice, but it decreases the cultivation of yam. It is observed that farmers respond to positive impact of climate on yields of cassava, maize, sorghum and rice by reallocating more land towards the cultivation of these crops. This is the overriding consideration in crop planting decisions in many developing economies. Sale of food stuffs to supplement family income is considered a secondary issue. It is therefore no wonder that food crop farmers respond weakly to price incentives, as they do not increase the size of farms for most of the crops in question in respond to positive change in output prices. Growers of cassava, maize and sorghum, however, react to changes in prices of related crops by putting more land into cultivation. This peculiar trait of food crop farmers can stifle the future development of the food crop subsector in Ghana.

The interpretation of the findings is conducted with the assumption that the only variables which will change in the future are climate variables. All other independent variables and even variables not considered in this study including population and technology are assumed constant over the projection period. The study also assumes that farmers can adapt fully to warming and dying by switching from one crop to another. However, this may not be the case because adaptation requires specific investments before it can be accessed.

Author details
[1]Debt Management Division, Ministry of Finance, P.O. Box MB 40 Accra, Ghana. [2]Graduate School of International Development and Cooperation, Hiroshima University, 1-5-1 Kagamiyama, Higashi Hiroshima 739-8529, Japan.

References

Branca G, McCarthy N, Lipper L, Jolejole MC (2013) Food security, climate change and sustainable land management: A review. Agronomy for sustainable development 33:635–650, doi: 10.1007/s13593-013-0133-1

Breisinger B, Diao X, Thurlow J (2009) Modeling growth options and structural change to reach middle income country Status: The Case of Ghana. Economic Modeling 26:514–525

Butt TA, McCarl BA, Angerer JA, Dyke PA, Stuth JW (2005) The Economic and food security implications of climate change in Mali. Climatic change 68(3):355–378

Cabas J, Weersink A, Olale E (2010) Crop yield response to economic, site and climate variables. Climate Change 101:599–616

Chipanshi AC, Chanda R, Totolo O (2003) Vulnerability assessment of the maize and sorghum crops to climate change in Botswana. Climatic change 61(3):339–360

Corobov R (2002) Estimations of climate change impacts on crop production in the Republic of Moldova. Geo journal 57:195–202

Darwin R, Tsigas M, Lewandrowski J, Raneses A (1995) World agriculture and climate change: Economic Adaptations. Agricultural economic report, Natural Resources and Environment Division, Economic Research Service, U.S. Department of Agriculture Agricultural Economic Report No. 703.

Garrity D, Akinnifesi F, Ajayi O, Weldesemayat S, Mowo J, Kalinganire A, Larwanou M, Bayala J (2010) Evergreen agriculture: a robust approach to sustainable food security in Africa. Food Security 2:197–214

GEPA (Ghana Environmental Protection Agency) (2001) Ghana's initial national communications report. United Nations Framework Convention on Climate Change, Accra

Hajivassiliou VA, McFadden D, Ruud P (1996) Simulation of multivariate normal rectangle probabilities and their derivatives: theoretical and computational results. Journal of Econometrics 72(1–2):85–134

Just R, Pope R (1978) Stochastic specification of production functions and economic applications. Journal of Econometrics 7:67–86

Kaiser HM, Riha SJ, Wilks DS, Rossiter DG, Sampath R (1993) A farm-level analysis of economic and agronomic impacts of gradual climate warming. American journal of agricultural economics 75(2):387–398

Knox JW, Hess TM, Daccache A, Wheeler T (2012) Climate change impact on food crop productivity in Africa and South Asia. Environmental Research Letters 7 034032

Kurukulasuriya P, Mendelsohn R (2006) Crop selection: adapting to climate change in Africa, CEEPA Discussion paper No. 18, Center for Environmental Economics and Policy in Africa, University of Pretoria.

Mendelsohn R, Dinar A (2009) Climate change and agriculture: An economic analysis of global impacts, adaptation and distributional effects. Edward Elgar Publishing Ltd., Cheltenham, UK and Northampton, MA, USA

Mendelsohn R, Nordhaus WD, Shaw D (1994) The impact of global warming on agriculture: a ricardian analysis. The American economic review 84(4):753–771

MOFA (Ministry of Agriculture) (2007) District level agricultural production and price data. Statistics Research and Information Directorate, Accra

Nhemachena C, Hassan RM (2007) Micro-level analysis of farmers' adaptation to climate change in Southern Africa. IFPRI Discussion paper 00714. International Food Policy Research Institute, Washington, IFPRI Discussion paper 00714

Ramteke AA, Shirgave PD (2012) Study the effect of common fertilizers on plant growth parameters of some vegetable plants. Journal of Plant Product and Natural Resources 2(2):328–333

Roodman D (2011) Estimating fully observed recursive mixed-process models with cmp. Stata Journal 11(2):159–206

Sagoe R (2006) Climate change and root crop production in Ghana. Retrieved June 8, 2012, from http://www.nlcap.net/fileadmin/NCAP/Countries/Ghana/ROOT_TUBERS_DRAFT_FINAL_REPORT.pdf

Seo SN, Mendelsohn R (2008) An analysis of crop choice: adapting to climate change in South American Farms. Ecological economics 67(1):109–116

Zellner A (1962) An efficient method of estimating seemingly unrelated regressions and tests for aggregation bias. Journal of American Statistical Association 57(298):348–368

Farm technology adoption in Kenya: a simultaneous estimation of inorganic fertilizer and improved maize variety adoption decisions

Maurice J Ogada[1*], Germano Mwabu[2†] and Diana Muchai[3†]

* Correspondence:
ogadajuma@yahoo.co.uk
[†]Equal contributors
[1]International Livestock Research
Institute (ILRI), P.O. Box
30709-00100, Nairobi, Kenya
Full list of author information is
available at the end of the article

Abstract

This paper models inorganic fertilizer and improved maize varieties adoption as joint decisions. Controlling for household, plot-level, institutional and other factors, the study found that household adoption decisions on inorganic fertilizer and improved maize varieties were inter-dependent. Other factors found to influence the adoption of the two technologies were farmer characteristics, plot-level factors and market imperfections such as limited access to credit and input markets, and production risks. Thus, easing market imperfections is a pre-requisite for accelerating farm technology adoption among the smallholders. Inter-dependence of farm technologies must also not be ignored in farm technology adoption promotion initiatives.

Keywords: Technology adoption; Simultaneous estimation; Africa; Kenya

JEL Classification: Q10; Q16; O55

Background

The Green Revolution which dramatically boosted the yield of cereals in Asia and Latin America is a clear manifestation of the potential of agricultural technologies in improving people's lives especially in the developing world (Pray, 1981). Indeed, it is the basis of support for Green Revolution in Africa by such philanthropic organizations as the Rockefeller and the Gates foundations. Successful agricultural transformation, the World over, has been largely attributed to improved farm technologies such as fertilizer, improved seeds and soil and water conservation (Johnston and Kilby, 1975; Mellor, 1976; Gabre-Madhin and Johnston 2002). Adoption of these technologies provides opportunities for increasing not only agricultural productivity but also incomes (Feder *et al.*, 1985). For developing countries, the contribution of improved technologies to agricultural productivity is well documented (see Sunding and Zilberman, 2001; and Doss, 2006 for details).

With the support of development partners, the government of Kenya has introduced or implemented several efficiency and productivity-enhancing technologies, programmes and projects at household level. Among the projects and programmes are the Kenya Agricultural Productivity Project (KAPP), the National Agriculture and Livestock Extension Programme (NALEP), the Agriculture Sector Programme Support (ASPS) and the National Accelerated Agricultural Inputs Access Programme (NAAIAP). Improved technologies for soil and water conservation, improved storage facilities, labour-saving and improved seeds have also been developed and disseminated, particularly by the Kenya Agricultural Research Institute (KARI).

Despite the efforts by the government and development partners, levels of technology adoption remain low (Republic of Kenya 2007; Ogada et al. 2010). While average adoption rates of improved maize varieties and inorganic fertilizer of 65 per cent and 76 per cent, respectively, appear impressive, great variations exist across regions and agro-ecological zones. The adoption rates are as low as 12 per cent for fertilizer (Olwande et al. 2009) and 30 per cent for improved maize varieties (Alliance for a Green Revolution in Africa 2010) in some regions. They are even worse, hardly 10 per cent, for other improved seed varieties (Alliance for a Green Revolution in Africa AGRA 2010). Given the link between technology adoption and farm productivity, and the desire by the Government of Kenya to promote development and adoption of agricultural technologies (Republic of Kenya 2007), understanding the factors that influence adoption of new and/improved technologies across households and communities is of urgent interest.

Previous studies have treated improved maize varieties and inorganic fertilizer as independent technologies, adopted independently (see Makokha *et al.*, 2001; Ouma *et al.*, 2002; Wekesa *et al.*, 2003; Olwande et al. 2009; and Ogada et al. 2010). If simultaneity in decision-making exists, this approach yields biased, inefficient and inconsistent estimates (Maddala, 1983; Greene, 2003). This paper overcomes this problem by employing bivariate model which captures the inter-dependence of the two decisions. Various factors, which are not necessarily obvious to researchers, can simultaneously affect adoption decisions between improved maize varieties and inorganic fertilizer. For instance, the two could have synergies in farm production. As a result, the farmer who adopts an improved maize variety could most likely adopt inorganic fertilizer. What this implies is that the farm households could be adopting technologies as a package, say improved maize variety with complementary element as inorganic fertilizer and pesticides (Kabila et al. 2000). Besides methodological novelty, this paper incorporates GIS-generated measures of location and space in the analysis unlike the previous studies which relied on use of dummy variables.

The remainder of the paper is organized as follows: section 2 explores empirical literature on improved farm technology adoption; section 3 discusses the methods employed; section 4 presents and discusses the results; and section 5 concludes and infers policy implications.

Brief review of literature

Empirical works on determinants of agricultural technology adoption have mainly focused on risk and uncertainty (Koundouri et al. 2006; Simtowe *et al.*, 2006), asymmetry of information, institutional constraints, human capital, access to inputs (Feder et al. 1985; Foster and Rosenzweig, 1996; Kohli and Singh, 1997) and availability of supportive infrastructure as possible predictors of adoption decisions. More recently, however, focus has extended to social networks and learning. The literature is vast and may not easily be compressed. Therefore, this section reviews only a few relevant studies starting with those from other developing countries before moving to Kenya.

Kohli and Singh (1997) conducted a comparative study on adoption of high yielding varieties (HYVs) among states in India and concluded that rapid adoption of the HYVs in Punjab was as a result of cheap and easy access to the technology itself and the complementary inputs. As noted by McGuirk and Mundlak (1991) in their study in India using choice of technique framework, HYVs require high levels of fertilizer input and irrigation to realize the yield potential. Thus, complementary inputs must be available and affordable to enhance adoption of HYVs.

Another strand of literature focuses on social networks and learning. The basic argument here is that adoption of technologies is influenced by Bayesian learning. That is, initially only a few farmers may adopt, and even this group of farmers may do this just on smaller experimental scales. As the first harvest is realized, the farmers can update their belief about the technology which may increase the rate of adoption in subsequent years. Besley and Case (1993), for instance, used a model of learning in a situation where profitability of technology adoption was uncertain and beyond the farmer's control in India. They found that probability of adopting an agricultural technology increases as farmers realize the profitability of the new technology. Using a target-input model of new technology which assumes that the best use of an input is unknown and random, Foster and Rosenzweig (1995), and Conley and Udry (2002) found similar results. Foster and Rosenzweig (1995) studied adoption of HYVs in India while Conley and Udry (2002) studied application of fertilizer in pineapple cultivation in Ghana. These authors concluded that initial adoption may be low due to imperfect information on management and profitability of the new technology but as this becomes clearer from the experiences of their neighbours and their own experiences, adoption is scaled up. This is supported by Bandiera and Rasul (2006) who examined the link between social networks and technology adoption in Northern Mozambique and noted that a farmer who discussed agriculture with others had a higher propensity to adopt new technologies.

In Kenya, Gerhart (1975) examined adoption of hybrid maize between late 1964 and 1973 in western Kenya and noticed a rapid diffusion of the technology in the region despite constraints. Risk and uncertainty were identified as the greatest constraints to adoption of hybrid maize. Factors found to encourage adoption decision were farmer education, access to credit, and access to extension services. Farmers who adopted hybrid maize were found more likely to adopt other yield-enhancing practices such as use of inorganic and organic fertilizers, and modern management practices such as planting in rows, weeding more than once, thinning and using insecticides. Gerhart applied multivariate probit on cross-sectional data and supplemented it with qualitative techniques. The approach was theoretically sound because technology adoption decisions are inter-dependent and the decision to adopt one technology could enhance or deter adoption of other related technologies. However, the use of cross-sectional data ignores the dynamic aspects of household adoption behaviour which could make the work less suitable for policy. Moreover, the study examined only one region of the country.

Other studies in the country have followed the path of the seminal paper by Griliches (1957) on adoption of new agricultural technologies. Griliches examined heterogeneity of local conditions and how it affected adoption of hybrid corn in the mid-western United States. He noted the role of economic factors such as expected profits in influencing the variation in farm technology spread rates. He further noted that speed of adoption across geographical locations depended on the suppliers of the technology and suitability of the seed to local conditions. It is indeed from the work of Griliches that economic literature on agricultural technology adoption developed. Some of the factors that possibly explain the rate of adoption and the long-run equilibrium level of use of new agricultural technology as identified in the economic literature include: credit constraints, risk aversion, the farmer's landholding size, land tenure system, human capital endowment, quality and quantity of farm equipment, and supply of complementary inputs (Feder et al. 1985). Among the studies that have adopted this approach are Makokha et al. (2001), Ouma et al. (2002), and Wekesa et al. (2003). Makokha et al. (2001) examined determinant of adoption of fertilizer and manure in Kiambu District, focusing on soil quality as reported by the farmers. They found high cost of labour and other inputs, unavailability of demanded packages and untimely delivery as the main constraints to fertilizer adoption. Ouma et al. (2002) focused on adoption of fertilizer and hybrid seed in Embu District and found that agro-climate, manure use, cost of hired labour, gender of the farmer and access to extension services were important determinants of adoption. Wekesa et al. (2003) examined adoption of improved maize varieties and fertilizer in the coastal lowlands of Kenya and found that unsuitable climatic conditions, high cost and unavailability of seed, perceived soil fertility and low financial endowments were responsible for the low adoption. The above findings are consistent with those of the International Maize and Wheat Improvement Center (CIMMYT) studies as summarized by Doss (2007). Other cross-sectional studies, though focusing on different technologies such as dairy and soil and water conservation, have found similar results (see Nicholson et al., 1999; Ogada et al. 2010; Oostendorp and Zaal, 2011). These studies have three main limitations: they are based on cross-sectional data, they cover smaller geographical areas that cannot accurately reflect the diversity among farming communities and they use ordinary binary probit or logit which ignores the inter-dependence of agricultural technologies. Their results are, thus, likely to suffer endogeneity bias.

A study by Olwande et al. (2009)) used panel data to examine determinants of fertilizer adoption and intensity of use. Using a double-hurdle model, they found that age and education of the farmer, access to credit, presence of a cash crop, distance to fertilizer market and agro-ecological potential influence the probability of fertilizer adoption. Gender of the farmer, dependency ratio, credit access, presence of cash crop, distance to extension services and agro-ecological potential were found to influence intensity of fertilizer use. A double-hurdle model is useful in capturing intensity of adoption but it ignores the fact that adoption of fertilizer could also be influenced by related practices such as adoption of improved maize seed.

Overall, literature indicates that household demographic, location, socio-economic and institutional factors are important determinants of farm technology adoption and equilibrium level of use. But their effects may not be universal. A factor is universal if it promotes or impedes technology adoption irrespective of location and type of technology. Rubas (2004) tested the universality of age, education, outreach, and farm size in influencing adoption of agricultural technologies. Employing Ordinary Least Squares (OLS) and Minimum Absolute Deviation (MAD) approaches, the study found that education and farm size were positive universal (encourage adoption of all types of technology irrespective of location) while outreach was not universal. Age was not universal by OLS, and negative universal by MAD. The fact that universality of age, education and farm size was confirmed to be weak, one cannot assume that farm households in different locations respond to different technologies in the same way.

Methods

Theoretical model

Because households in Kenya and elsewhere in developing countries produce under uncertainty and great market imperfections, the study adopted expected utility maximization framework. Production risk was represented by the stochastic term, ε, whose distribution, $G(.)$ was exogenous to the farm household's actions. Inorganic fertilizer and improved maize varieties being among the most important inputs in the smallholder's crop production process, the household's production function was presented as:

$$q_{it} = q\left(X_{it}^f, X_{it}^s, X_{it}^o, \varepsilon_{it}\right) \tag{1}$$

where X_{it}^f and X_{it}^s represent fertilizer and seed inputs by the household in a given year, respectively. X_{it}^O represent other inputs while ε_{it} represent production risk. This function was assumed to be well-behaved. Hereafter, the panel dimensions are suppressed for simplicity.

Letting r and p represent input and output prices respectively, the problem of a risk-averse household is to maximize expected utility of gross income expressed as:

$$\underset{X}{Max}\, E[U(\pi)] = \underset{X}{Max} \int_0^q \left[U\left(pq(X_f, X_s, X_o, \varepsilon) - r_f X_f - r_s X_s - r_o X_o\right)\right] dG(\varepsilon) \tag{2}$$

Where $U(.)$ is the Von Neumann-Morgenstern utility function. E is the expectation operator while ε captures all the unobserved household heterogeneity such as unreported farm management ability, land fertility, risk preferences and risk management measures, and rate of discount which could affect input use and farm productivity.

Given that r and p are non-random, the first order necessary condition (FONC) for the fertilizer variable was specified as:

$$E\left(r_f U'\right) = E\left[p\frac{\partial q(X_f, X_s, X_o, \varepsilon)}{\partial X_f} U'\right] \tag{3}$$

And

$$\frac{r_f}{p} = E\left[\frac{\partial q(X_f, X_s, X_o, \varepsilon)}{\partial X_f}\right] + \frac{COV\left(U'; \partial q(X_f, X_s, X_o, \varepsilon)/\partial X_f\right)}{E(U')} \tag{4}$$

where U' is the change in utility of income due to change in income. That is, $\frac{\partial U(\pi)}{\partial \pi}$. FONC for the other variables were derived using the same procedure. For risk-neutral households, the second term on the right hand side of Equation 4 would be equal to zero and therefore adoption of improved farm technologies would be dependent on the traditional marginal conditions. For the risk-averse households, the term would be different from zero and would measure deviations from the risk neutrality position. The term would be proportional and opposite in sign to the marginal risk premium with respect to the input under consideration (Koundouri et al., 2006). In such a case, adoption of improved farm technology would be influenced by production risk besides the cost of technology adoption and farm-specific factors that may influence either technology performance or adoption costs.

Market imperfections made it important to include household characteristics and resource endowments in explaining farm household's investment and production decisions (Pender and Kerr, 1998). For example, labour market imperfections constrain a household's labour demand to its own labour supply with the result that only larger households are able to invest in labour-intensive technologies. Similarly, capital market imperfections restrict households to their savings and already accumulated capital assets such that poorer households are not able to invest in capital-intensive technologies. Generally, a household invests in a given improved farm technology if the expected utility with adoption, $E[U(\pi_{wa})]$, is higher than expected utility without adoption, $E[U(\pi_{-wa})]$. That is, when $E[U(\pi_{wa})] > E[U(\pi_{-wa})]$.

Antle (1983; 1987) provide a flexible way to estimate Equations 3 and 4 which only requires information on prices, input quantities and other observable variables. The approach equates maximizing expected utility of farm income with respect to any input to maximizing a function of moments of the distribution of ε. The moments themselves have X_f and X_s as arguments (Antle, 1983; 1987). This study, thus, computed the first three moments of the stochastic production function and included them as covariates in analysing adoption decisions for inorganic fertilizer and improved maize seed.

The study hypothesized that adoption decisions by a household on improved maize (HM_{ha}) and inorganic fertilizer ($fert_{ha}$) were interdependent. The decisions also depended on the profitability of the technology (P_f) as measured by proximity to market and access road, land ownership system (L_{os}), access to credit and market (A_{cm}), and household information on the improved technologies ($Info_h$). Other factors were plot characteristics (P_c), household characteristics (h_c), agro-ecological characteristics (AE_c), production risk (P_r) and other random factors (see Pender and Kerr, 1998; Shiferaw and Holden, 1999; Doss, 2007; Yesuf and Kohlin, 2008). The theoretical model of improved maize variety and inorganic fertilizer adoption decisions was, thus, specified as follows:

$$fert_{ha} = f\left(HM_{ha}, p_f, L_{os}, A_{cm}, Info_h, P_c, h_c, AE_c, P_r, \mu_{fert}\right) \tag{5}$$

$$HM_{ha} = f\left(fert_{ha}, P_f, L_{os}, A_{cm}, Info_h, P_c, h_c, AE_c, P_r, \mu_{HM}\right) \tag{6}$$

Equations 5 and 6 represent the observed binary variables which reflect the latent net benefits, $fert_{ha}^*$ and HM_{ha}^*, from adopting inorganic fertilizer and improved maize variety, respectively.

Empirical model

As specified in the theoretical model (Equations 5 and 6), production risks are important in household decisions to adopt improved maize varieties and inorganic fertilizer. This is because farm households in low income countries are risk averse (Dercon, 2004). They suffer welfare erosion when their consumption and production fluctuate. This fluctuation may be captured by yield variability. However, it would be wrong to assume that the variance of production captures all the production risks to which households are exposed (Di Falco and Chavas, 2009). For instance, households are exposed to the downside risk (the risk of crop failure as measured by the skewness, with negative skewness implying greater exposure to crop failure). Analysis of adoption decisions was, therefore, done in two steps. In the first step, the first three sample moments of the maize yield distribution (mean, variance and skewness) were computed. In the second step, the estimated moments were included in the adoption models together with other explanatory variables.

The first step involved regressing maize yield in a given year against plot, household and village-level variables to obtain estimates of mean effects. The general functional form of the model was:

$$q_{it} = q(X_{it}, Z_{it}, \beta_1) + e_{it} \tag{7}$$

In estimating the yield-response function (Equation 7), we applied Guan *et al.* (2006) approach. This approach breaks the production function into two parts: the crop growth function and the scaling function. The crop growth function is specified as quadratic function instead of translog, commonly used in literature. Quadratic function is preferred to translog because it permits concavity and zero input. Concave yield-response curves are indeed consistent with most observable biological relationships. For instance, excess fertilizer or rainfall adversely affects crop growth. The scaling function incorporates facilitating inputs such as crop management practices, government programmes and household socio-economic attributes. This part of the production function is specified in exponential form which does not impose monotonicity on input–output relationship (Guan *et al.*, 2006).

The crop production model estimated, therefore, had the following general functional form:

$$y_{it} = G(X).F(Z) \tag{8}$$

where y_{it} is the maize yield realized by a household in a particular crop year, X is a vector of growth inputs, Z is a vector of facilitating inputs, G(.) is the crop growth function while F(.) is the facilitating or scaling function. Essentially, the crop growth function determines the attainable yield level given the biophysical environment while the interaction between the growth factors and the facilitating factors determines the actual yield. This explains why even farmers operating under similar agro-ecological conditions experience yield differences.

The quadratic crop growth part was specified as:

$$G_{it} = \alpha_1 rain_{it} + \alpha_2 plantingfert_{it} + \alpha_3 dressfert_{it} + \alpha_4 manure_{it} + \alpha_5 rain_{it}^2 + \\ \alpha_6 rain_{it} plantingfert_{it} + \alpha_7 rain_{it} dressfert_{it} + \alpha_8 rain_{it} manure_{it} + \\ \alpha_9 plantingfert_{it}^2 + \alpha_{10} plantingfert_{it} dressfert_{it} + \alpha_{11} plantingfert_{it} manure_{it} + \\ \alpha_{12} dressfert_{it}^2 + \alpha_{13} dressfert_{it} manure_{it} + \alpha_{14} manure_{it}^2 \tag{9}$$

The scaling function was specified as:

$$F_{it} = \exp\{-[\beta_0 + \beta_1 acres_{it} + \beta_2 flbmpha_{it} + \beta_3 flbfpha_{it} + \beta_4 flbcpha_{it} + \\ \beta_5 headeducyrs_{it} + \beta_6 soccaplev + \beta_7 crpinc_{it} + \beta_8 mktdist]^2\} \tag{10}$$

Overall, the maize yield function was expressed as:

$$maizpha_{it} = \{\alpha_1 rain_{it} + \alpha_2 plantingfert_{it} + \alpha_3 dressfert_{it} + \alpha_4 manure_{it} + \alpha_5 rain_{it}^2 + \\ \alpha_6 rain_{it} plantingfert_{it} + \alpha_7 rain_{it} dressfert_{it} + \alpha_8 rain_{it} manure_{it} + \alpha_9 plantingfert_{it}^2 + \\ \alpha_{10} plantingfert_{it} dressfert_{it} + \alpha_{11} plantingfert_{it} manure_{it} + \alpha_{12} dressfert_{it}^2 + \\ \alpha_{13} dressfert_{it} manure_{it} + \alpha_{14} manure_{it}^2\} \exp\{-\beta_0 + \beta_1 acres_{it} + \beta_2 flbmpha_{it} + \\ \beta_3 flbfpha_{it} + \beta_4 flbcpha_{it} + \beta_5 headeducyrs_{it} + \beta_6 soccaplev + \beta_7 crpinc_{it} + \\ \beta_8 mktdist]^2\} + f_i + e_{it} \tag{11}$$

where f_i refers to unobserved household heterogeneity and e_{it} s the random error term assumed to be i.i.d.N(0, σ^2). All the other variables are defined in Additional file 1: appendix 1. $\alpha_1, \alpha_2,......., \alpha_{14}$ and $\beta_0, \beta_1,...., \beta_8$ are the parameters to be estimated.

The jth central moment of the maize yield about its mean was, therefore, computed as:

$$e_j = e\left\{ [q(.)-\mu]^j \right\} for\ j\ =\ 2, ...,\ m \tag{12}$$

where μ is the mean maize yield or the first moment of maize yield per household. The estimated residuals from the mean regression were the estimates of the first moment of maize yield distribution. The estimates were then squared and regressed against the same variables as in Equation 13.

$$e_{it}^2 = q_2\left(X_{it}, Z_{it}, \hat{\beta}_2\right) + v_{it} \tag{13}$$

The least squares estimates of $\hat{\beta}_2$ are consistent and asymptotically normal (Antle, 1983). The predicted values of e_{it}^2 are also consistent estimates of the second central moment of maize production distribution. The same procedure was used to estimate the third central moment (skewness) of maize production distribution (in this case the estimated errors were raised to power three). In literature, this approach has been used by Antle (1983) and Koundouri $et\ al.$ (2006).

The estimated production risk factors were then incorporated into the improved maize variety and inorganic fertilizer adoption models (Equations 14 and 15) which were estimated as bivariate probit to deal with simultaneity of technology adoption decisions. Similar approach has been used by Feder $et\ al.$ (1985), Feder and Onchan (1987) and Yesuf and Kohlin (2008). The model was specified as follows:

$$fert_{ha}^* = \gamma_1 HM_{ha} + \alpha_1' X_1 + \varepsilon_f \ ,\ fert_{ha} = 1 \begin{cases} if\ fert_{ha}^* > 0 \\ \\ 0\ otherwise \end{cases} \tag{14}$$

$$HM_{ha}^* = \gamma_2 fert_{ha} + \alpha_2' X_2 + \varepsilon_H \ ,\ HM_{ha} = 1 \begin{cases} if\ HM_{ha}^* > 0 \\ \\ 0\ otherwise \end{cases} \tag{15}$$

$$\{\varepsilon_f, \varepsilon_H\} \sim BVN\left\{(0,0), \sigma_f^2, \sigma_H^2, \rho\right\},$$

Where P is the correlation, σ_i^2 is the variance,$fert_{ha}^*$ and HM_{ha}^* are observed binary (latent) variables indicating the household's adoption status of fertilizer and improved maize. X_1 and X_2 are vectors of explanatory variables (including production risk factors), and ε_f and ε_H are the error terms for the respective equations.

The reduced form of the model required for consistent and efficient estimates was:

$$fert_{ha} = \pi_1' X + e_f \tag{16}$$

$$HM_{ha} = \pi_2' X + e_H, \tag{17}$$

$$\{e_f, e_H\} \sim BVN\left\{(0,0), \sigma_f^2, \sigma_H^2, \tau\right\},$$

where τ is the correlation (measure of the extent to which the two errors covary), σ_i^2 is the variance, and X is the union of exogenous variables in the system. The correlation coefficient between the errors measures the extent of correlation between inorganic fertilizer and improved maize varieties adoption decisions. This arises if unobserved variables that affect adoption of inorganic fertilizer also affect adoption of the improved maize varieties. In the presence of such correlation, univariate probit yields biased esti-mates while bivariate probit technique produces consistent and fully efficient estimates (Greene, 1998). Bivariate probit model is estimated by maximum likelihood.

Coefficients of bivariate probit, just like those of other discrete choice models cannot be interpreted directly. They are transformed into marginal effects, interpreted as the change in predicted probability associated with the changes in the exogenous variables. Following Greene (1998), the marginal effects were computed as:

$$\frac{\partial BVN\left[\Phi\left(a_1'X_1 + \gamma_1, a_2'X_2 + \gamma_2, \rho\right)\right]}{\partial z_k} = \left\{\phi\left(a'X_1 + \gamma_1\right)\Phi\left[\left(\gamma_2'X_2 - \rho\left(a_1'X_1 + \gamma_1\right)\right)/\right.\right.$$
$$\left.\left.\sqrt{1-\rho^2}\right]\right\}a_z + \left\{\phi\left(\gamma_2'X_2\right)\Phi\left[\left(\left(a_1'X_1 + \gamma_1\right) - \rho\left(\gamma_2'X_2\right)\right)/\sqrt{1-\rho^2}\right]\right\}\gamma_z,$$

for fertilizer, and:

$$\frac{\partial BVN\left[\Phi\left(a_2'X_2 + \gamma_2, a_1'X_1 + \gamma_1, \rho\right)\right]}{\partial z_k} = \left\{\phi\left(a'X_2 + \gamma_2\right)\Phi\left[\left(\gamma_1'X_1 - \rho\left(a_2'X_2 + \gamma_2\right)\right)/\sqrt{1-\rho^2}\right]\right\}a_z +$$
$$\left\{\phi(\gamma_1'X_1)\Phi\left[\left((a_2'X_2 + \gamma_2) - \rho(\gamma_1'X_1)\right)/\sqrt{1-\rho^2}\right]\right\}\gamma_z, \text{ for improved maize varieties. } \Phi \text{ is}$$
the normal cumulative distribution function. When $p = 0$ the expression reduces to:
$$\Phi(a_2'X_2)\Phi(a_1'X_1 + \gamma_1) + \Phi(-\gamma_2'X_2)\Phi(a_1'z_1).$$

Estimation challenges and remedies

For consistent estimates of production risks and determinants of adoption of improved maize varieties and inorganic fertilizer, it was important to control for unobserved hetero-geneity (f_i) which might have been correlated with observed explanatory variables. Using household fixed effects could have been an option because household panel data were available. Unfortunately, the Guan et al. (2006) approach used to estimate production risks and the bivariate probit model used to estimate adoption are non-linear maximum likeli-hood models which cannot be directly estimated by fixed effects (Wooldridge, 2002). As a result, the study adopted Mundlak (1978) and Chamberlain (1984) approach. The approach involved including mean values of time-varying explanatory variables in Equations 11, 16 and 17. That is, the approach assumes that unobserved effects are linearly correlated with explanatory variables as expressed in Equation 18.

$$f_i = \tau + \gamma\bar{X}_i + \eta_i \tag{18}$$

where $\bar{X}_i$ is a vector of the mean of time-varying explanatory variables within each household, τ is a constant, γ is a vector of parameters and $\eta_i \sim \text{i.i.d}(0, \sigma_\eta^2)$ and is inde-pendent of e_{it}, e_f and e_H . The vector γ is not different from zero if the observed ex-planatory variables are uncorrelated with the random effects.

Data

The study used the Tegemeo Agricultural Monitoring and Policy Analysis (TAMPA) household panel survey data. The survey is a collaboration project between Tegemeo Institute of Egerton University, Kenya, and Michigan State University of the United States. It aims at monitoring smallholder production patterns, consumption and incomes to identify policy agenda for farmers. It is geographically diverse and nationally representative of the rural maize-growing areas. The panel is based on the Kenya National Bureau of Statistics' (KNBS) agricultural sample frame. Only two waves of the survey, 2004 and 2007, were available to us with complete information on 1167 farm households. The waves contain detailed information on agricultural input and output, household consumption, income, demographics, location, infrastructure and credit information.

Summary statistics of the variables for bivariate analysis of technology adoption are provided in Tables 1 and 2. Adoption rates by survey years and broad agro-ecological zones (lowlands, midlands and highlands) are first examined (Table 1).

Adoption rates of inorganic fertilizer and improved maize varieties increased between the two survey periods for all the broad agro-ecological zones. However, the lowlands remained the lowest adopters of both technologies, recording only 3.8 per cent adoption rate for inorganic fertilizer and 38 per cent for improved maize varieties. For manure adoption, the highlands were the worst performers although this was appropriately compensated for by the high adoption of inorganic fertilizer. Notably, adoption rates of manure did not change between the periods of reference. The 3.8 per cent of households that adopted inorganic fertilizer in the lowlands in 2004 also adopted improved maize varieties. The same pattern was replicated in 2007 with 3.9 per cent of households adopting both technologies, an indication that out of the 7.7 per cent that adopted inorganic fertilizer most also adopted improved maize varieties. This provides a useful insight: those who adopt inorganic fertilizer are more likely to adopt improved maize varieties. Descriptive statistics of the other variables used in the analysis are provided in Table 2.

The adopting households exhibited certain characteristics over their non-adopter counterparts for the two technologies. They had higher levels education, and marginally higher average age. A larger proportion of these households were male headed, had soils with poor water-retention, and had more land under maize cultivation. On average, such households were also closer to input markets. On the production risks, these households experienced higher expected maize yield, higher yield variability and higher downside risk. In the 2004 survey, the adopting households reported a lower wage rate for farm labour than the non-adopter households. In 2007, however, reported wage rates were, on average, similar for the two categories of households.

Results and discussion

Estimation of the production function was useful only for generating production risks. The results are shown in Additional file 1: appendix 2. Focus of this section, therefore,

Table 1 Household adoption levels of fertilizer and improved maize varieties

Technology	2004			2007		
	LL	ML	HL	LL	ML	HL
I/fertilizer	3.8	60	86	7.7	63	89
Maize	38	60	83	39	65	88
Both	3.8	48	73	3.9	49.7	79.4
Manure	45	44	25	46	46	25
No. of observations	1167			1167		

LL = Lowland; ML = Midlands; HL = Highlands; I/fertilizer = Inorganic fertilizer

Table 2 Descriptive statistics of bivariate probit analysis variables

Variable	2004				2007			
	I/Fertilizer		I/Maize		I/Fertilizer		I/Maize	
	NA	A	NA	A	NA	A	NA	A
Age of household head (years)	52 (23)	54 (17)	53 (23)	53 (18)	52 (25)	54 (22)	50 (27)	54 (21)
Household head's education (0 = No education; 1 = primary level; 2 = secondary level; 3 = tertiary level)/fraction of households in each category								
0	0.37	0.18	0.36	0.19	0.36	0.23	0.40	0.22
1	0.46	0.49	0.48	0.48	0.47	0.46	0.43	0.48
2	0.15	0.26	0.13	0.26	0.15	0.24	0.14	0.24
3	0.02	0.07	0.03	0.07	0.02	0.07	0.03	0.06
Household size (No. of people in a household)	4 (2)	4 (2)	4 (2)	4 (2)	5 (3)	5 (3)	5 (3)	5 (3)
Acres (size of cropped land in acres)	1.4 (1.6)	1.5 (1.9)	1.3 (1.6)	1.5 (1.9)	1.2 (1.4)	1.8 (3.8)	1.2 (1.3)	1.7 (3.7)
Male-headed households (Proportion of male-headed household)	0.73 (0.44)	0.84 (0.37)	0.73 (0.44)	0.84 (0.37)	0.70 (0.46)	0.79 (0.41)	0.66 (0.47)	0.80 (0.40)
Credit access (proportion of households that received credit)	0.26 (0.09)	0.27 (0.11)	0.27 (0.09)	0.26 (0.11)	0.26 (0.08)	0.27 (0.11)	0.27 (0.10)	0.26 (0.11)
Market distance (distance from household to input market in kilometres)	7.2 (8.2)	5.6 (6)	6.7 (7.8)	6.0 (6.5)	7.3 (8.4)	5.6 (5.9)	6.4 (7.8)	6.2 (6.6)
Soil water retention (ability of soils to retain water. Captured through GIS)	0.7 (0.46)	0.88 (0.32)	0.74 (0.44)	0.85 (0.36)	0.69 (0.46)	0.88 (0.33)	0.74 (0.44)	0.84 (0.36)
Expected yield (mean maize yield)	407 (159)	943 (392)	510 (310)	857 (413)	517 (135)	1,002 (390)	590 (250)	936 (405)
Yield variance (maize yield variability as predicted from the production function)	259,796 (308,208)	457,116 (723,916)	285,004 (488,765)	432,982 (652,834)	289,302 (309,176)	415,752.5 (572,709)	274,432 (302,657)	415,070 (560,788)

Table 2 Descriptive statistics of bivariate probit analysis variables (*Continued*)

Downside risk (Skewness of the maize yield as predicted from the production function)	$3.59D^{08}$ ($1.88D^{09}$)	$6.44D^{08}$ ($4.13D^{09}$)	$3.48D^{08}$ ($2.71D^{09}$)	$6.37D^{08}$ ($3.76D^{09}$)	$6.55D^{08}$ ($1.86D^{09}$)	$4.39D^{08}$ ($3.35D^{09}$)	$4.18D^{08}$ ($1.70D^{09}$)	$5.68D^{08}$ ($3.32D^{09}$)
Wage rate (Daily wage rate of the farm labour in Kenya Shillings)	87 (40)	82 (32)	85 (41)	83 (32)	89 (37)	89 (29)	89 (36)	89 (30)
Number of observations	1167			1167				

Standard deviations in parentheses, I/Fertilizer = Inorganic fertilizer; I/Maize = Improved maize varieties; NA = Non-Adopters; A = Adopters.

is the results of the bivariate probit model of adoption of inorganic fertilizer and improved maize varieties. A joint significance test of the average terms rejected the null hypothesis, $H_0 : \gamma = 0$ (Equation 18) for the production function and the adoption Equations. This meant that the unobserved heterogeneity was correlated with the averages, $\bar{X}_i$. Mundlak-Chamberlain approach was, therefore, justified.

For all the models, the problem of multicollinearity was tested and found to be serious for variance and skewness (reflected in the variance inflation factor of 13.27 and 11.09, respectively). To solve the problem, skewness was dropped from the analysis and the test re-conducted. The test combined Variance Inflation Factor (VIF) and Eigen Values approaches. All the variance inflation factors were less than 2 and the condition number was 2.66, indicating that multicollinearity was no longer a problem.

Bivariate probit results are displayed in Table 3. The significance of LR test ($\rho = 0$) implies that adoption decisions about improved maize varieties and inorganic fertilizers are not independent. Both decisions are affected by the same unobservable heterogeneities. Thus, the decisions are jointly determined. This is plausible because fertilizer and improved maize varieties are complementary agricultural production technologies. Therefore, estimation of separate equations yields unreliable results. The finding is consistent with those of McGuirk and Mundlak (1991) and Kohli and Singh (1997).

The smallholders using manure in their crop production were six per cent less likely to adopt inorganic fertilizer. Such households were, however, seven per cent more likely to adopt improved maize varieties. Inorganic fertilizer and manure would be complementary in circumstances where there is under application of both but basically the two should be substitutes. Thus, farm households that have and apply sufficient quantities of manure would not apply inorganic fertilizer. This explains why both increase the probability of adopting improved maize varieties. For joint adoption of inorganic fertilizer and improved maize varieties, manure adopting smallholders had a six per cent lower chance than their non-adopting counterparts.

Education of the farm household's head was important in influencing joint adoption of the two technologies under consideration. Households whose heads had primary school level and secondary school level of education had four per cent and five per cent higher chance, respectively, than their uneducated counterparts. Positive correlation between education and technology adoption was also noted by Gerhart (1975). However, as indicated by Rubas (2004), universality of education in influencing technology adoption, though positive, is weak. Thus, it is not surprising that its influence is statistically insignificant for adoption of inorganic fertilizer and improved maize variety singly but significant for their joint adoption.

While the gender of the household head had no effect on adoption of the individual technologies, it weakly promoted joint adoption of the two technologies. Male-headed households had four per cent higher probability of adopting both inorganic fertilizer and improved maize variety than the female-headed households. This possibly indicates that female-headed households are more resource-constrained.

Increased regional access to agricultural credit is important in promoting adoption of inorganic fertilizer and joint adoption of inorganic fertilizer and improved maize variety. Improvement of credit access index by one per cent, improves the probability of households adopting inorganic fertilizer by 26 per cent and joint adoption of inorganic fertilizer and improved maize variety by 20 per cent. This is consistent with the findings

Table 3 Determinants of inorganic fertilizer and improved maize varieties adoption decisions

Variable	Inorganic fertilizer adoption		Improved maize varieties		Joint adoption
	Coefficient	Marginal effect	Coefficient	Marginal effect	Marginal effect
Technology Adoption					
Manure	−0.79*** (-6.04)	−0.06***(-3.68)	0.03 (0.31)	0.07*** (4.23)	−0.06***(-3.53)
Human and physical capital					
Head's age	−0.02** (-2.47)	0.001 (1.34)	0.001 (0.13)	−0.001 (-1.30)	−0.001(-1.03)
Head's age Sq	0.0002**(2.49)		−0.0001(-0.96)		
Head's education					
Primary	0.1 (0.75)	−0.02 (-1.28)	0.19**(1.97)	0.02(1.15)	0.04*(1.93)
Secondary	0.14 (0.84)	−0.03 (-1.53)	0.28**(2.21)	0.03(1.32)	0.05**(2.10)
Tertiary	0.18 (0.73)	−0.02 (-0.71)	0.24(1.29)	0.02(0.55)	0.05(1.48)
Male head	0.18 (1.24)	−0.01 (-0.70)	0.18 (1.61)	0.01(0.53)	0.04*(1.92)
Household size	−0.02 (-0.88)	−0.0001(-0.02)	−0.01 (-0.51)	0.0005 (0.13)	−0.003(-0.93)
Wage rate	−0.0002 (-0.09)	−0.00001 (-0.04)	0.0003 (0.16)	−0.0002 (-0.10)	−0.00004(-0.13)
Institutional factors					
Credit access	3.1*** (4.85)	0.26*** (4.14)	−0.39 (-0.92)	−0.31*** (-4.60)	0.20**(2.23)
Secure tenure	−0.08 (-0.97)	−0.04***(-4.18)	0.26*** (3.94)	0.04*** (4.08)	0.03**(2.33)
Market distance	−0.04*** (-3.05)	−0.001(-0.52)	−0.01** (-2.17)	0.001(0.99)	−0.01***(-3.42)
Plot and soil characteristics					
Fast-drained soils	0.86*** (6.75)	0.02(1.38)	0.29*** (3.12)	−0.03 (-1.58)	0.11***(6.44)
Plot Size (acres)	0.39*** (6.89)	0.01(0.79)	0.14*** (2.75)	−0.01(-1.59)	0.05***(5.87)
Plot size (acres) Sq.	−0.004***(-4.49)		0.0004(0.18)		
Year 2007	−0.14 (-1.55)	−0.03***(-2.83)	0.14**(2.27)	0.03***(2.82)	0.01 (0.74)
Production Risks					
Expected yield	0.004*** (11.60)	0.0002*** (6.64)	0.001*** (3.24)	−0.0002*** (-7.90)	0.0004***(11.45)
Variance	$-3.63D^{-07}$*** (-2.57)	1.32D (0.01)	$-1.73D^{-07}$* (-1.94)	$6.01D^{-09}$ (0.46)	−5.47D-08***(-2.73)
Means of time-varying variables					
Mean head age	0.002 (0.33)		0.002(0.40)		
Mean acres	−0.26*** (-4.48)		−0.12**(-2.34)		
Mean wage	−0.002 (-0.58)		0.003 (1.43)		
Mean household size	0.03 (0.76)		0.017 (0.62)		
Mean expected yield	0.002*** (6.26)		0.002***(7.59)		
Mean variance	$-4.13D^{-07}$*(-1.79)		$5.96D^{-08}$(0.45)		
Intercept	−4.41*** (-12.64)		−1.87*** (-8.27)		
LR test ($\rho = 0$)	$\chi^2(1) = 19.9$***		$\chi^2(1) = 19.9$***		
No. of Observations	2334		2334		

***, ** and * indicate significance at 1%, 5% and 10%, respectively; figures in parentheses are Z-scores.

of Feder et al. (1985) and Olwande et al. (2009) Smallholders may not be able to accumulate sufficient savings to purchase relatively more expensive technologies like inorganic fertilizer or combined inorganic fertilizer and improved maize variety. On the contrary, increased credit access lowers the probability of adoption of improved maize variety as an individual technology. This implies that access to credit could make smallholders switch to higher value crops.

Land tenure security is important in influencing adoption of improved maize variety and joint adoption of inorganic fertilizer and improved maize variety. Households with secure land tenure had four per cent higher probability of adopting improved maize variety and three per cent higher chance of adopting combined inorganic fertilizer and improved maize variety than their counterparts with insecure land tenure regime. While it is not explicit from our data, it is possible that secure tenure enables households to lease out part of their landholding for some cash for purchase of the improved technologies.

Distance to input market was negatively correlated with joint adoption of inorganic fertilizer and improved maize variety. A household which is one kilometre closer to the input market had one per cent higher chance of adopting both inorganic fertilizer and improved maize variety than its counterpart one kilometre away. Most probably this is due to easier access to these technologies by farm households closer to the markets. Households located far from markets essentially incur higher costs of adoption due to transport charges.

Households whose plots were well-drained had 11 per cent higher chance of joint adoption of inorganic fertilizer and improved maize varieties than households with poorly drained plots. Well-drained soils are highly vulnerable to erosion and leaching. This could substantially reduce their fertility, increasing the need to adopt improved technologies to enhance output. This is consistent with Wekesa et al. (2003).

The size of plot cultivated by the household was positively correlated with joint adoption of the two technologies. An increase of a household's cultivated land area by one acre, on average, increased the probability of joint adoption of inorganic fertilizer and improved maize varieties by five per cent. Literature attributes positive influence of plot size on improved technology adoption to confounding factors such as poor soil quality, fixed costs of adoption, credit access and risk preferences (Feder et al. 1985). This study controlled for the confounding factors but plot size was still significant in positively influencing probability of adoption of the two technologies. This supports the Neo-Malthusian hypothesis that land redistribution and fragmentation arising from population pressure does not lead to more intensification of farming.

While time had influence on the adoption of the individual technologies, it had no effect on joint adoption of the two technologies. Relative to 2004, the 2007 adoption of inorganic fertilizer was three per cent lower. The reverse was true of improved maize variety adoption.

Expected higher yield enhanced probability of adoption of inorganic fertilizer and joint adoption of inorganic fertilizer and improved maize variety. On the contrary, highly variable yield lowered probability of joint adoption. This indicates that smallholders are risk averse and would be hesitant to invest in highly uncertain activities. Negative influence of risk and uncertainty on farm technology adoption has previously been noted by Gerhart (1975), Koundouri et al. (2006) and Simtowe et al. (2006).

Conclusion and policy implications

Stagnating agricultural productivity has been a major policy concern in Kenya. It has led to increased investment in development and dissemination of yield-enhancing technologies. Remarkable success has been recorded in adoption of inorganic fertilizers and improved maize varieties although wide disparities remain across geographical areas. For other improved crop varieties, adoption levels remain very low, barely 10 per cent of the farm households. Thus, this study sought to understand the drivers of adoption of improved farm technologies among the smallholder food crop farmers in the country. It examined bivariate adoption of inorganic fertilizer and improved maize varieties to control for unobservable household heterogeneities in adoption decisions.

The study found that decisions to adopt complementary technologies are interdependent. It further established that plot-level, household-specific factors, and market imperfection are important in influencing the likelihood of a household adopting inorganic fertilizer and improved maize varieties. Among the key factors in this regard include education level of the household head, plot size operated by the household, land tenure security, distance to the input market, water-retaining capacity of the plot, access to credit, manure adoption, expected yields and yield variability.

The above results have important policy ramifications. Foremost, it is important to consider the complementarity of different agricultural technologies in promotion of their adoption. For instance, smallholders may be hesitant to adopt improved maize varieties if they are unable to obtain fertilizer to go with it. Thus, to promote adoption of complementary technologies, it is important to ensure that the technologies are available and affordable to the smallholders. For example, it may not be useful to subsidize one of the technologies without due consideration of the famers' capability to fully fund the remaining parts of the cost of adoption.

Although larger plots attract adoption of inorganic fertilizer and improved maize varieties, it may not be possible to curtail further sub-division of agricultural land as population increases. One option could be to increase access to land through land rental market to enable land-constrained smallholders acquire additional farmland. This is possible through land banks. Another option, though achievable only in the long term, is to expand the industrial sector to absorb more people from the agricultural sector to reduce pressure on agricultural land.

Improved technologies should be availed within easy reach of the farming households. While the government can contribute to this by improving transport infrastructure within the farming villages, the technology producers and marketers have the most important role of setting up distribution outlets closer to the farming communities. Local farmer organizations may also contribute through bulk buying of the improved technologies and directly supplying the same to the members in appropriate quantities.

To deal with the influence of yield and yield variability on farm technology adoption, it is important to ensure that the yield-enhancing technologies are able to increase yields substantially and maintain the high yields. Thus, when a technology is associated with high risks that may lead to extreme yield fluctuations, it may be useful to insure the farmers against such risks to encourage adoption. Index-based crop insurance is an option that could be explored.

Setting up smallholder credit scheme, especially for purchase of farm technologies, could be an important step towards accelerating farm technology adoption. Because the smallholders may not be able to acquire credit from the mainstream financial sector due to the risky nature of their business, the government could step in either as a guarantor or as a direct provider of the funds through, say microfinance institutions. An alternative approach could be to mobilize the smallholders to form organizations through which to pool resources and obtain additional funding from either the government or financial institutions. Whichever approach is chosen, the funds should be low-interest and easily accessible.

The above policy implications are short-run remedial measures. Long-run solutions, however, lie in correcting market imperfections. This is only possible with broad-based economic development.

Additional file

Additional file 1: Appendices.

Competing interests
The authors declare that they have no competing interests.

Authors' contributions
MJO carried out econometric analysis and drafted the results chapter. GM conceptualized the study and undertook literature review. DM undertook literature review and data analysis. All the authors read and approved the final manuscript.

Author details
[1]International Livestock Research Institute (ILRI), P.O. Box 30709-00100, Nairobi, Kenya. [2]School of Economics, University of Nairobi, P.O. Box 30179-00100, Nairobi, Kenya. [3]School of Economics, Kenyatta University, P.O. Box 43844-00100, Nairobi, Kenya.

References
Alliance for a Green Revolution in Africa (AGRA) (2010) Baseline Study, Kenya Report. Tegemeo Institute of Agricultural Policy and development, Nairobi

Antle JM (1983) Testing for stochastic structure of production: a flexible moment-based approach. Journal of Business and Economic Statistics 1(3):192–201

Antle J (1987) Econometric estimation of producers' risk attitudes. American Journal of Agricultural Economics 69(3):509–522

Bandiera O, Rasul I (2006) Social networks and technology adoption in northern mozambique. Economic Journal 116(514):869–902

Besley T, Case A (1993) Modelling technology adoption in developing countries. American Economic Review 83:396–402

Chamberlain G (1984) Panel Data. In: Griliches Z, Intriligator MD (eds) Handbook of Econometrics, 2nd edn. North Holland, Amsterdam, pp 1247–1318

Conley T, Udry CR (2002) Learning about a new technology: pineapple in Ghana, working paper, Yale University

Dercon S (2004) Growth and shocks: evidence from rural Ethiopia. Journal of Development Economics 74(2):309–329

Di Falco S, Chavas JP (2009) On crop biodiversity, risk exposure, and food security in the highlands of Ethiopia. American Journal of Agricultural Economics 91(3):599–611

Doss CR (2006) Analyzing technology adoption using microstudies: limitations, challenges, and opportunities for improvement. Agricultural Economics 34:207–219

Doss CR (2007) Understanding Farm-Level Technology Adoption: Lessons Learned from CIMMYT's Micro-Surveys in Eastern Africa. Economics Working Paper, Nairobi pp 03–07, CIMMYT

Feder G, Onchan T (1987) Land ownership security and farm investment in Thailand. American Journal of Agricultural Economics 69:311–20

Feder G, Just RE, Zilberman D (1985) Adoption of Agricultural Innovations in Developing Countries: A Survey. Economic Development and Cultural Change 33(2):255–98

Foster A, Rosenzweig M (1995) Learning by doing and learning from others: human capital and farm household change in agriculture. Journal of Political Economy 103(6):1176–1209

Foster AD, Rosenzweig MR (1996) Technical change and human-capital returns and investments: Evidence from the green revolution. The American Economic Review 86(4):931–53

Gabre-Madhin E, Johnston B (2002) Accelerating Africa's Structural Transformation: Lessons from Asia. In: Jayne TS, Isaac M, Gem A-K (eds) Perspectives on Agricultural Transformation: A View from Africa. Nova Science, New York

Gerhart JD (1975) The Diffusion of Hybrid Maize in Western Kenya, Ph.D. Dissertation, Princeton University

Greene W (1998) Gender economics courses in liberal arts colleges: further results. Journal of Economic Education 29:291–300

Greene W (2003) Econometric Analysis, 5th edn. New York University, New York

Griliches Z (1957) Hybrid corn: an exploration in the economics of technological change. Econometrica 25(4):

Guan Z, Lansink AO, Van Ittersum M, Wossink A (2006) Integrating agronomic principles into production function estimation: a dichotomy of growth inputs and facilitating inputs. American Journal of Agricultural Economics 88:203–214

Johnston BF, Kilby P (1975) Agriculture and Structural Transformation: Economic Strategies in Late-Developing Countries. Oxford University Press, New York

Kabila A, Verkuijl H, Mwangi W (2000) Factors affecting adoption of improved seeds and use of inorganic fertilizer for maize production in the intermediate and lowland zones of Tanzania. Journal of Agricultural and Applied Economics 32(1):35–47

Kohli DS, Singh N (1995) The Green Revolution in Punjab, India: The Economics of Technological Change. Revised version of the paper presented at a conference on Agriculture of Punjab at the Southern Asian Institute, Columbia University

Koundouri P, Nauges C, Tzouvelekas V (2006) Technology adoption under production uncertainty: theory and application to irrigation technology. American Journal of Agricultural Economics 88(3):657–670

Maddala GS (1983) Limited-Dependent and Qualitative Variables in Econometrics. Cambridge University Press, New York

Makokha S, Kimani S, Mwangi W, Verkuijl H, Musembi F (2001) Determinants of Fertilizer and Manure Use for Maize Production in Kiambu District, Kenya. CIMMYT (International Maize and Wheat Improvement Center), Mexico, D.F

McGuirk AM, Mundlak Y (1991) Incentives and Constraints in the Transformation of Punjab Agriculture, IFPRI Research Paper No. 87. International Food Policy Research Institute (IFPRI), Washington, D.C

Mellor JW (1976) The New Economics of Growth: A Strategy for India and the Developing World. Cornell University Press, Ithaca, NY

Mundlak Y (1978) On the pooling of time series and cross-section data. Econometrica 46:69–85

Nicholson CF, Thornton PK, Mohamed L, Muinga RF, Mwamachi DM, Elbasha EH, Staal SJ, Thorpe W (1999) Smallholder dairy technology in coastal Kenya. In: Sechrest L, Stewart M, Stickle T (eds) A synthesis of findings concerning CGIAR case studies on the adoption of technological innovations. IAEG Secretariat, FAO, Rome

Ogada MJ, Nyangena W, Yesuf M (2010) Production risk and farm technology adoption in rain-fed semi-arid lands of Kenya. AfJARE 4(2):

Olwande J, Sikei G, Mathenge M (2009) Agricultural Technology Adoption: A Panel Analysis of Smallholder Farmers' Fertilizer use in Kenya. CEGA Working Paper Series No. AfD-0908. Centre of Evaluation for Global Action. University of California, Berkeley

Oostendorp RH, Zaal F (2011) Understanding adoption of soil and water conservation techniques: The role of new owners. CSAE Working Paper. Department of Economics, Oxford University, Oxford, United Kingdom, WPS/2011-05

Ouma J, Murithi F, Mwangi W, Verkuijl H, Gethi M, De Groote H (2002) Adoption of Maize Seed and Fertilizer Technologies in Embu District, Kenya. CIMMYT (International Maize and Wheat Improvement Center), Mexico, D.F

Pender JL, Kerr JM (1998) Determinants of farmers' indigenous soil and water conservation investments in semi-arid India. Agricultural Economics 19:113–125

Pray C (1981) The Green Revolution as a Case Study in Transfer of Technology. Annals of the American Academy of Political and Social Science Nov 458:68–80

Republic of Kenya (2007) Kenya Vision 2030, Government of the Republic of Kenya, Ministry of Planning and National development and the National Economic and Social Council (NESC), Office of the President

Rubas D (2004) Technology Adoption: Who is likely to adopt and how does timing affect the benefits? PhD Thesis, Texas A&M University

Shiferaw B, Holden S (1999) Soil Erosion and Smallholders: conservation Decisions in the highlands of Ethiopia. World Development 27(4):739–752

Simtowe FJ, Mduma AP, Alban T, Zeller M (2006) Can risk-aversion towards fertilizer explain part of the non-adoption puzzle for hybrid maize? empirical evidence from Malawi. Journal of Applied Sciences 6(7):1490–1498

Sunding DL, Zilberman D (2001) The Agricultural Innovation Process: Research and Technology Adoption in a Changing Agricultural Sector. In: Gardner B, Rausser G (eds) Handbook of Agricultural and Resource Economics. Elsevier, North Holland, Amsterdam, p 2001

Wekesa E, Mwangi W, Verkuijl H, Danda VK, De Groote H (2003) Adoption of Maize Technologies in the Coastal Lowlands of Kenya. CIMMYT, Mexico, D.F.

Wooldridge JM (2002) Econometric Analysis of Cross Section and Panel Data. The MIT Press, Cambridge, Massachusetts, USA

Yesuf M, Kohlin G (2008) Market Imperfections and Farm Technology Adoption Decisions: A case study from the Highlands of Ethiopia. EfD DP 08–04, RFF, Washington D.C

Evaluating willingness to become a food education volunteer among urban residents in Japan: toward a participatory food policy

Yasuo Ohe[*], Shinichi Kurihara and Shinpei Shimoura

* Correspondence:
yohe@faculty.chiba-u.jp
Department of Food and Resource
Economics, Chiba University, 648
Matsudo, Matsudo, Chiba 271-8510,
Japan

Abstract

Food education has attracted growing concern focusing on appropriate dietary habits amid the deterioration of many aspects of health in this industrial society. For this purpose, a self-motivating type of food policy framework is expected to increase in significance. Under such a self-motivating policy framework that aims at inducing people to increase their food-conscious behaviour from the aspects of a healthy diet, food safety, reduction of food waste, and the local community, this paper investigated the awareness of food education by a survey of 3,000 inhabitants in a suburb of Tokyo and explored the factors determining the willingness to participate as a food education volunteer. We found the existence of a generation gap in terms of interest in and knowledge regarding food education, which suggests that there is room for working together to exert a complementary role between generations.

JEL classification: D62; D64; I18; Q18; Q58

Keywords: Food education; Volunteer education; Community awareness; Dietary behaviour; Educational externality; Environmental awareness; Health awareness

Background

The issue of food education has attracted growing concern regarding appropriate dietary habits in our industrial society. For that purpose, it is necessary to promote self-motivated behaviour to raise the awareness of various aspects of food-related behaviour. The policy framework that stimulates self-motivating behaviour will play an increasingly important role as civil society progresses toward a greater number self-motivated citizens. In this context, the Basic Law on Food Education was established in Japan to take advantage of this trend. After a long discussion of whether the state should inter-vene in personal life, the Basic Law on Food Education was enacted in 2005 in Japan and aims to enhance national awareness of matters related to food and nutrition and to eventually improve behaviours related to food and dietary aspects in daily life Maff (2010). Basically, although behaviours associated with food and health are personal matters in which the state should not intervene, it has been said that self-motivating measures that aim to enhance health awareness have been increasing in importance amid the deterioration of many aspects of health in every industrial society. This is a common issue facing these countries.

The focus of education on food differs from one country to another due to differences in issues of health, food security, and dietary heritage, including local foods and communal integrity because food has a connection with many aspects of life (McMichael 2000). Therefore, food education involves a wide range of generations, from children to the elderly, and also involves a wide range of concerns such as the awareness of various aspects of food, health, the environment, and the local community. In this regard, food education is not limited to education in schools, but includes the realm of adult education, which tries to transform the framework of behaviour of adults over time (Mezirow 1997; 2003). Nevertheless, in the arena of agricultural economics, food education issues have not been fully addressed despite intensive studies on food consumption and issues of food safety.

Education has externality because education of the public enhances economic efficiency, which benefits the whole society (Arai 1995). Food education has the same effect. This is because an increase in food awareness improves people's health and results in a reduction in medical costs and food waste, which eventually benefits the entire society through better resource allocation. A self-motivating policy framework, however, often ends up with merely setting goals without any concrete results. In this respect, to ensure the effectiveness of a self-motivating framework it is necessary to enlarge the domain of self-motivating behaviour. This is also necessary from the aspects of financial constraints that the public sector commonly faces and of the maturation of a civil society. Thus, the authors believe that it is necessary to explore the factors that determine the willingness of people to take part in volunteering to offer food education in connection with attainment of policy goals under a self-motivating policy framework. Although care should be taken not to expect too much from volunteers, in order to reduce administrative costs and promote the progress of a self-motivated civil society this point should be clarified. Although a simple questionnaire survey was conducted to better understand the national awareness of food education issues by central and prefectural governments at the time of enactment of the law (Cabinet Office 2007a; Kanagawa Prefecture 2007) and studies regarding specific food-related issues have been conducted in many countries, to our knowledge, there has been no full-scale research on the awareness of food education and willingness of individuals to volunteer to offer food education.

In this paper, those who volunteer are understood to be people who undertake economic behaviour that intentionally accompanies externality. We approach the issue by the following steps. First, based on a literature review, we present conceptual and empirical models for volunteer behaviour. Second, based on a questionnaire survey of 3,000 residents in Matsudo, an urban municipality that comprises an extended metropolitan area, we quantitatively explored what and how inhabitants' attitudes are connected with daily food-related behaviour and also estimated a willingness determinant model for participation as a food education volunteer taking into account the complementary role among generations. Finally, we give policy recommendations for a more participatory way of conducting a food education program.

Literature review

In reviewing the literature on food education topics that covered a wide range of aspects, we could find very few studies on food education and volunteering per se although food choices and food and nutrition policies have been widely examined from economic viewpoints (Evenson 1981; Blaylock et al. 1999; Kenkel and Manning 1999).

Studies that we reviewed dealt with aspects related to food education and conscious-ness of food itself, such as food safety, food and health, waste of food and environmen-tal issues regarding food in relation to the community.

First, as for food education issues, Edwards and Hartwell (2002) examined primary school children's attitudes toward and knowledge of fruits and vegetables from the viewpoint of a balanced diet. Pivarnik et al. (2009) statistically examined knowledge of food safety and food safety education among high school teachers and supported the need for a food safety education program. Roe and Teisl (2007) investigated consumer reaction to labeling of genetically modified foods. Heslop et al. (2007) examined the be-haviour of single working mothers in performing food-related household tasks. Scott et al. (2009) performed a statistical examination of outcomes of a safety education program for consumers on practices in using fresh produce and concluded that the food safety education program was effective in teaching consumers. von Normann (2009) described the increasing importance of food education at school based on results of an empirical examination of the impact on food patterns in German children based on lifestyle and knowledge of food. Lautenschlager and Smith (2007) examined the effects of community gardens on dietary behaviour of youth and found that garden programs had a positive impact on food choices, knowledge of nutrition and cooking skills among youth.

In relation to the aspect of environmental and community issues, Chappell and LaValle (2011) focused on the relationship between food security and biodiversity and argued that the two things are not mutually exclusive by using alternative agricultural practices. Griffin et al. (2009) estimated food waste across an entire local community. DeLind (2011) explored the effectiveness of contextual analysis of local food movements. Kerton and Sinclair (2010) focused on the significance of consumer participation in local organic agriculture and the learning experienced through that participation as a way of transformative learning advocated by Mezirow (1997, 2003). Hughner et al. (2007) pointed out that demographics and beliefs of organic food consumers were more heterogeneous than generally perceived. Kemp et al. (2010) examined the 'food miles' concept from the revealed preference survey of UK consumers and found that consumers respected this concept much less than expected. Iyer and Kashyap (2007) explored the positive condi-tions for consumer participatory recycling and noted that the provision of incentives and information was effective in increasing participation.

Now, turning to health issues, McGinnis and Meyers (1999) discussed nutrition and health policy by focusing on the Dietary Guidelines for Americans and drew on broad policy interventions. Robinson and Smith (2003) investigated the connection between health conscious consumers, body mass index (BMI), and attitudes about sustainably produced food and found that health conscious consumers were more likely to be female, older, more educated, earn a higher income, be more active, and have a healthier BMI than those who were not health conscious consumers. Cash et al. (2005) investigated possible health benefits of "thin subsidies", consumption subsidies for healthier foods, in the United States. Walsh and Nelson (2010) found the continuing influence of parents on adolescents' dietary behaviour by investigating the link between diet and health from a questionnaire survey of adolescents in Northern Ireland and recommended practical nutritional education for balanced dietary behaviour. Chang and Nayga (2011) highlighted the connection between mothers' nutritional label use and children's obesity and noted that mothers' nutritional label use leads to a lower probability of children becoming overweight. Visschers et al.

(2013) conducted a cluster analysis of nutrition information usage and health and nutrition interests and identified four segments with different demographics. Twiss et al. (2003) focused on the role of community gardens to enhance public health in improving the quality of life, which is supported by local leadership and volunteers. This topic is also connected with a community building effect.

Second, we could not find studies on volunteers who provide food education, so we reviewed studies on volunteers in general, which seem to be classified roughly into two categories: motives or recruitment issues and issues of effects. First, with respect to motives, Allison et al. (2002) statistically examined motives of volunteers in the community by comparison of two kinds of data sources and predicted the frequency of volunteering. Bussell and Forbes (2001) mentioned that those who are well educated, female and over the age of 50 are more likely to volunteer. Callow (2004) explored the motives for volunteering among retirees and found that the heterogeneous nature of senior citizens should be considered for volunteer recruitment. Thus, the profile of pro-volunteer citizens has been clarified, although, on the other hand, heterogeneity of the pool of volunteers should be borne in mind. Freeman (1997) revealed that the standard explanation of the labour supply for volunteering accounted for a minor part of volunteer behaviour and that volunteering is a 'conscience good', which is something that people feel morally obligated to do when asked, but which they would just as soon let someone else do. Handy et al. (2000) developed the net-cost theory that hypothesized that the public perception of volunteering will be based on the perception of the net cost incurred by the individual and empirically verified this hypothesis from a cross-cultural perspective. This net-cost theory provided suggestions in building our conceptual framework below.

As to studies that evaluated the effects of volunteering, Brown (1999) assessed the dollar value of volunteer activity in the U.S and concluded that the overall value was understated when the gains accruing to volunteers themselves were included. Handy and Srinivasan (2004) evaluated the economic value of the net benefits of hospital volunteers and found a quite high value for every dollar spent. Hager and Brudney (2005) explored the measure of net benefit that takes into account both benefits and challenges of a volunteer program. Mook et al. (2005) examined an accounting paradigm that considers the value of volunteer contributions. Bowman (2009) applied microeconomic theory to the evaluation of volunteering and pointed out that the positive externalities that volunteers generate to society as a whole were not well calculated. In short, we can say that evaluation of the economic value of volunteers is worthwhile although an exact evaluation of the entire value of volunteering is difficult due to positive externalities.

With respect to food education in a Japanese context, Adachi (2008) urged the necessity for nutrition education from a holistic perspective on food and nutrition. Mah (2010) examined food education in Japan from the perspective of decentralized public health governance. Kimura (2011) discussed issues of food education in Japan from the viewpoint of privatized and gendered food knowledge. Ohe (2012) examined the educational function of agriculture. Earlier Ohe (2011) evaluated the educational function of dairy farming with an economic framework as a process of internalization of a positive externality.

To summarize, there is extensive literature mentioning that participatory and educational programs are effective tools to raise food-related consciousness among people. Nevertheless, it should be noted that no study dealt with the willingness to become a food education volunteer either at the conceptual or empirical level from an economic point of view.

Methods

Conceptual framework

Here, we set up an economic framework for the subsequent empirical investigation based on the net-cost theory that assumed that volunteers would incur the net cost of volunteering. Figure 1 illustrates two cases of a resident's subjective equilibrium on educated food behaviour, measuring value vertically and the level of educated food behaviour horizontally. The more rightward, the better food awareness behaviour is undertaken. The optimal level of food awareness behaviour is determined by the two curves: marginal benefit gained from that behaviour, MB, and marginal cost necessary for that behaviour, MC. We do not assume the social cost for that behaviour. As ordinarily expected, the MB curve decreases and the MC curve increases when the level of behaviour increases. With respect to the marginal benefit, there are two curves depicted in Figure 1 because food conscious behaviour exerts a positive educational externality on society. The upper marginal cost curve represents the social marginal benefit curve, MSB, while the lower curve represents the private marginal benefit curve, MPB, and the vertical gap between the two curves demonstrates the externality, which is the benefit to society due to diffusion of educated food behaviour.

If there is no externality at all, a resident's behaviour is determined at the private optimal point e_p where the MPB curve meets the MC curve. If externality exists, the optimal point e_s is determined where the MSB curve meets the MC curve, which is the social optimal point. Thus, the socially optimal level of educated food behaviour is conducted at os. In this case, attaining the social optimal externality should be internalized. Put differently, generators of externality should be compensated for the amount of externality. Otherwise, externality is not generated because it is rational

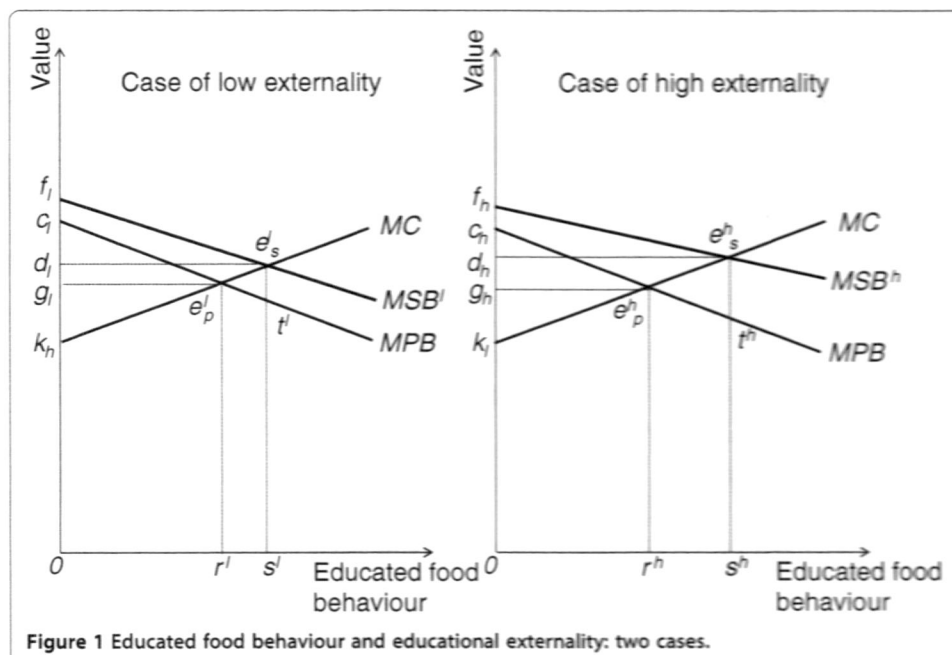

Figure 1 Educated food behaviour and educational externality: two cases.

for residents to undertake educated food behaviour at the private optimal point e_p in the long run if they are not compensated.

Now, let us consider the case of residents who conduct volunteer work. Volunteers are considered those who do not internalize the externality by compensation from outside, but by bearing the opportunity cost by themselves at the social optimal point. Thus, we can say that volunteers are those who bear the opportunity cost of volunteering that generates externality, which is depicted as $e_s t$ in Figure 1. Here, to simplify the discussion, we assume no public subsidy and that volunteers bear all of the opportunity cost by themselves. Thus, generation of externality and enlargement of self-sustaining food education effects are expected through volunteering activity.

Whether volunteering is actually undertaken or not depends on the comparison between the expected utility from volunteering and the opportunity cost for volunteering. In this sense, the educational externality represents not only the opportunity cost for volunteering but also the expected benefit generated to society from volunteering. Thus, it is safe to say that the expected marginal benefit to society, $e_s t$, will be included in the expected marginal utility for volunteering.

Nevertheless, it becomes a burden for people to bear that opportunity cost on their own, so it is quite natural that the capability to assume that burden varies from one to another. It is most likely that if an individual is better able to bear that cost, the externality will be larger as depicted in the right case in Figure 1, while if that individual is not, the externality will be smaller as illustrated in the left case. Thus, in the case of high externality educated food behaviour is more highly achieved than that of low externality because $os^h > os^l$. Empirically, the question to be clarified is what factors in reality make these differences.

Analytical model

Here, based on the conceptual framework the researchers present an analytical model for the following empirical estimation of the determinant function of the willingness to participate as a food education volunteer. Generally, volunteer activity is different from ordinary economic activities from the aspect that an income gain is not the prime cause of action.

Since volunteer activity is not income earning behaviour, the volunteer activity needs to satisfy the following two conditions: Equation (1) and Equation (2). First, since the majority of the respondents were female, we take into account household production activity. To simplify the discussion here, we assume that an opportunity cost for participation in the labour market and for household production are identical, so that participation in the labour market and household production are indifferent in terms of utility level. Then an ordinary urbanite decides his or her own behaviour when a given wage level, W, exceeds the urbanite's subjective reservation wage, wr, after which the urbanite decides to take part in the labour market to gain income or undertaking household production (Equation (1)).

$$W(\text{wage level}) \text{ is greater than } wr(\text{subjective reservation wage}) \tag{1}$$

Second, it is considered that the expected utility of volunteer participation, $EU(vf)$, is lower than the utility level of the reservation wage, $U(wr)$, as indicated in Equation (2).

U(*wr*)(utility level of reservation wage) is greater than   (2)
EU(*vf*)(expected utility of volunteer participation)

This is because if the expected utility level of volunteer participation is greater than the reservation wage, then the resident can be better off by joining the labour market. We assume that these two conditions are satisfied.

Given these conditions, if an urbanite's expected utility level gained from the volunteer activity, EU(*vf*), is greater than the opportunity cost of bearing the externality, U(*op*), then the resident will take part in that volunteer activity (Equation (3)).

In contrast, if EU(*vf*) is lower than U(*op*), then a resident will not take part in the volunteer activity (Equation (4)).

EU(*vf*) (expected utility of volunteer activity) is equal to or greater than   (3)

    U(*op*) (opportunity cost of bearing externality)

EU(*vf*) (expected utility of volunteer activity) is less than   (4)

    U(*op*) (opportunity cost of bearing externality)

Given the analytical framework on participation in volunteer activity above, we set up an empirical model, indicated as Equation (5), to identify influential factors that determine the willingness to participate as a food education volunteer. Since the areas of food education extend widely, specifically we consider the vector of household attributes, vector of food awareness, vector of environmental awareness, vector of community awareness, and health awareness to test multiple aspects of life. If parameters of these factors are positive, these factors raise the subjective expected utility level of volunteer activity and vice versa.

$$V = G(houshld, food, envrn, health, community, \varepsilon)$$ (5)

Where, *V* = willingness to participate as a food education volunteer (yes = 1, no = 0)
houshld = vector of household attributes
food = vector of food awareness
envrn = vector of environmental awareness
health = vector of health awareness
community = vector of community awareness
ε = stochastic error term

Let us explain these variables. Explained variable *V* is the willingness to participate as a food education volunteer (yes = 1, no = 0). The explanatory variables are those that are supposed to raise the expected utility of volunteer participation based on the results of the literature review. **houshld** is the vector of household attributes, which determines the opportunity cost for participation as a volunteer from household aspects. For example, whether or not the existence of childcare work stipulates this opportunity cost. **food** is the vector of food awareness, and it is reasonable to assume that the higher this awareness, the higher the willingness to volunteer.

Likewise, with **envrn**, i.e. the vector of environmental awareness, it is probable that the higher this awareness, the more positive the attitude toward becoming a food education volunteer from the instance that less trash emission is expected from a household with high environmental awareness. **health** is the vector of health awareness, and it is safe to say that the higher the health awareness among urbanites, the more careful they are on a daily basis about development of metabolic syndrome. As a result, those urbanites with higher health awareness will also have a higher willingness to participate as a volunteer. *community* is the vector that represents the degree of awareness of how residents themselves are connected with the local community. People with high community awareness are expected to be willing to volunteer.

Given the above hypotheses on the variables, the sign conditions for food awareness, environmental awareness, health awareness, and community awareness are expected to be positive while those of household factors are variable dependent. If these hypotheses are verified, we can say that the promotion of food education among urbanites will be achieved through the improvement in awareness of these three factors in their lives. Actual variables are determined after the statistical tests described below.

Results and discussion

Outline of basic law on food education, study area and data collection

The Basic Law on Food Education was established in 2005 after a long discussion over whether the state should intervene in the personal lives of citizens amid mounting concerns over food-related behaviours of consumers and conditions in domestic agriculture. Of concern was the declining food self-sufficiency in Japan despite the large amount of food wasted at the same time; the increase in lifestyle diseases, even among the young, that continuously raises the nation's medical costs; food safety; and loss of the traditional healthy dietary habits and local food heritage. Three ministries were responsible for food education policy: Cabinet Office, Ministry of Education and Science, and Ministry of Agriculture, Forestry and Fisheries (Cabinet Office 2006; 2007b). They established the nation's basic plan to promote food education to tackle those issues described above. Then every prefectural government was asked to set up its own basic promotion plan and municipalities were required to make an effort to set up their own basic promotion plan at the municipal level. This law refers, in Article 22, to the importance of the partnership between volunteers for food education and local governments (Cabinet Office 2006, p.123).

Our study area, Matsudo, which is in Chiba prefecture, has 480 thousand inhabitants and is a suburb of Tokyo prefecture, with only a 20-minute train ride to downtown Tokyo. From this favourable location, the majority of Matsudo residents commute to the Tokyo metropolitan area. Data were collected by a surface mail questionnaire that we conducted for those residents living in Matsudo in collaboration with the municipality of Matsudo, which is one of the most ardent municipalities with regard to food education in Chiba because of the growing concerns over the fast-paced urban lifestyle that emphasizes convenience among residents. Matsudo inaugurated the city's own basic promotion plan in 2008, which mentions that volunteers are among those stakeholders who are supposed to play a crucial role in food education and help by connecting with people to attain three main goals: firstly, having healthy dietary habits taking into account the importance of food; secondly, knowing more about local foodstuffs and how they are produced and distributed; and, thirdly, balancing food intake and maintaining good oral health for mastication to attain a mentally and physically healthy life (Matsudo City 2008, pp. 9–10, p.25).

After consideration of the results of two preceding questionnaire surveys on food education conducted by the state and Kanagawa prefecture (Cabinet Office 2007a;

Kanagawa Prefecture 2007), we asked 68 questions to elucidate the lifestyles and characteristics of the respondents such as basic knowledge regarding food education, food and related matters, perception toward environmental, health, and local community issues, and their attributes. These wide-ranging questions were presented because the area of food education includes not only dietary education itself, but also addresses issues related to the environment, health, and local community matters. Requests for information on income level, which would be a key indicator, and academic background were finally excluded as question items because of privacy considerations raised by city officials. The city officials worried about possible complaints from respondents claiming infringement of their privacy. After discussion with officials on the pros and cons, we finally dropped those questions to comply with the request from the officials. We established the survey sheet taking into account the opinions of the municipality officials and sent it to 3,000 residents over the age of 20 years who were randomly extracted from the basic resident registry. For households receiving the survey, we asked that those responding should be mainly those in charge of purchasing ingredients for meals and cooking. The survey period was from October 18th to October 31st, 2007. Of the 3,000 surveys, 1,262 were returned and 1,254 (41.8%) were used for examination after excluding questionnaires that did not provide information on the respondent's age, sex and willingness to be a food education volunteer.

Respondents' profile

Briefly, the profile of the respondents was as follows. First, the majority (81%) were female, which was not unexpected because we requested that responders be those mainly in charge of purchasing ingredients and cooking. Also, because of this, as to profession, full-time housewives or househusbands accounted for one third of the respondents, followed by company employees (24%) and part-time employees (21%); these three categories accounted for about 80% of respondents. Composition according to age group presented no large bias with 45% of responders being younger than 50 years of age and 55% of responders aged 50 or over. Nevertheless, only 14% were in their twenties, 36.8% were younger than 40 years of age, and 63.2% were 40 years old or over, which indicates that the majority of samples were middle-aged or past middle-age. Three fourths of respondents were married. Seventy percent had lived in Matsudo for more than ten years and over half (53%) had lived in this city for more than 20 years. Two-generation families, which are the typical nuclear family in modern society, accounted for 57% of respondents while traditional large families of three generations or more made up only about 6% of those responding; the remainder were married couples (29%) and those living alone (9%).

Statistical analysis of respondents' attributes with regard to food education

We employed statistical tests to identify the differences in awareness of food, food education, health, environmental issues, and involvement in local community issues.

Knowledge of and interest in food education

Firstly, Table 1 shows the results of survey items on food and food education issues. Health-related issues attracted the highest concern, followed by food safety and dietary habits while interest in food waste, the decline in the rate of food self-sufficiency and

Table 1 Concerns related to food issues and food education

Items Percentage of response (%)	Age group		Test results (sample size)	Sex		Test results (sample size)
	<40 years	≥40 years		Females	Males	
Healthy growth of mind and body	55.4	52.3	n(1254)	58.0	33.3	***(1254)
Lifestyle diseases	37.7	44.7	**(1254)	43.5	35.9	**(1254)
Food safety	33.6	45.6	***(1254)	43.3	31.6	***(1254)
Disorders of eating habits	40.5	33.6	**(1254)	37.3	30.7	*(1254)
Respect for food and nature	16.2	13.9	n(1254)	16.0	9.1	***(1254)
Increase in obesity and underweight	15.8	12.5	+(1254)	14.2	11.7	n(1254)
Food waste	14.3	12.5	n(1254)	13.3	12.6	n(1254)
Decline in food self-sufficiency	7.6	11.6	**(1254)	9.2	14.3	**(1254)
Gastronomic and local heritage	10.6	8.0	+(1254)	8.4	11.3	+(1254)
Environmentally friendly food production	3.0	6.1	**(1254)	4.4	7.4	*(1254)
Building relationship with farmers	0.2	1.5	**(1254)	1.1	0.9	n(1254)

Note: Chi-square test was employed. ***, **, *, + show 1%, 5%, 10%, 20% (as reference) significance levels, respectively, and n shows no significance. For each item, yes = 1, no = 0; no snswer also = 0.

local gastronomic culture and tradition were not high. Those aged 40 years or older showed interest in health issues such as lifestyle diseases, food safety, and disorders related to dietary habits. Female respondents had greater interest in mental and physical healthy growth of children and respect for food and nature while interest among males was greater in the rate of food self-sufficiency.

More than half of respondents answered that matters most important in food education for children were the reduction in food waste, respect for nature and food producers, and table manners (Table 2). Issues considered more or less important were food-related issues such as communication through eating, experiences in farming, forestry and fisheries, and succession of gastronomic heritage.

With respect to the site of food education, a significant difference in consensus was observed between age groups and sex, but respondents selected their home as the primary place for that purpose; more than half of those 40 years or older answered that their home was the preferred place while only a third of those who did so were under the age of 40 (Table 3). In contrast, the younger generation selected an equal role between home and school. A higher proportion of females than males responded that the home was the place for such education. In short, food awareness is higher among those 40 years old or older than among those under 40.

Eating habits and foodstuffs

Although the majority of respondents indicated that they ate breakfast every day, this ratio was higher in those 40 years old or older and in females with 1% significance (Table 4). 15.6% of male respondents said they rarely have breakfast every day. Thus, we can say that the younger generation and the male respondents eat breakfast less frequently.

That the yearly frequency of celebrations at home featuring traditional foods is low might be due to the loss of seasonality in urban life; this is in contrast to the results of the survey by the national government that covered not only urban areas, but also rural areas (Cabinet Office 2007a). We observed statistically significant differences between the two age groups examined; nearly 60% of those over 40 held celebrations featuring traditional foods more than 5 times each year in contrast to less than half under age

Table 2 Important items for food education of children

Item	Reduction in food waste	Respect for nature and food producers	Table manners	Better diet habbits	Communication through food issues	Experience in farm, forestry and fisheries	Succession of gastronomic heritage
Important	76.6	75.1	68.5	59.9	44.1	22.9	22.1
More or less important	22.0	23.9	28.6	37.7	50.2	62.2	64.3
Less important	1.4	0.9	2.8	2.4	5.7	14.0	12.7
Not important	0.0	0.0	0.0	0.0	0.0	0.9	0.9
Total (%)	100.0	100.0	100.0	100.0	100.0	100.0	100.0
Sample size	214	213	213	212	211	214	213

Note: Total might not be 100% because each item is rounded off.

40. Among families of two generations or more, 62.1% had such celebrations 5 times or more yearly while 42.7% of those living alone and married couples with no child did not.

In buying foodstuffs, food safety was the first concern of those age 40 or over followed in the order of nutrition, taste, and cost while those under 40 looked at costs due to the existence of children in the family. Likewise, those 40 years or older 'always' checked the origins of the food more than those under 40 (66.3% vs. 60.6% with 1% significance). These percentages were also higher for multi-generation families than for other types of families.

Understandably from the above differences, we observed differences in the level of knowledge of food safety between age groups and gender; the knowledge level was higher in those 40 years or older and in females compared with those under the age of 40 and males.

Health matters and environmental issues

As the first indicator of the degree of respondents' health awareness, we asked about self-awareness of metabolic syndrome, as was asked in the national survey. We found that those 40 years and older and male respondents indicated such awareness, with a significantly higher portion of those indicating that they had metabolic syndrome or had the potential to have it; 34.1% of those 40 years and over answered that they potentially had metabolic syndrome, as did 35.5% of the male respondents (Table 5).

Table 3 Site of food education

Place of food education	Age group		Sex	
	<40 years	≥40 years	Females	Males
Home	32.9	52.2	45.6	42.9
Somewhat home	30.5	22.4	26.8	19.1
Either home or school	34.2	21.3	25.2	29.9
Somewhat school	1.5	1.0	1.1	1.7
School	0.4	0.5	0.0	2.6
No answer	0.4	2.7	1.4	3.9
Total (%)	100.0	100.0	100.0	100.0
Test result (sample size)	***(1254)		***(1254)	

Note: Significance level is the same as in Table 1.

Table 4 Food awareness

Frequency of breakfast	Age group		Sex	
	<40 years	≥40 years	Females	Males
Everyday	74.9	89.4	86.8	72.9
Only on weekdays	5.8	1.8	3.1	3.9
2-3 times a week	9.1	4.3	5.5	8.7
Rarely	10.2	4.3	4.4	15.6
No answer	0.0	0.3	0.2	0.0
Total (%)	100.0	100.0	100.0	100.0
Test result (sample size)	***(1254)		***(1254)	

Frequency of celebration at home with traditional foods	Age group		Family composition	
	<40 years	≥40 years	Married couple or single	Two generations or more
Five times or more	47.4	58.5	42.7	62.1
Test result (sample size)	***(1254)		***(1254)	

Concern over food purchases	Age group		Sex	
	<40 years	≥40 years	Females	Males
Nutritional factors	18.2	19.2	19.5	16.0
Food safety	33.6	50.6	47.9	28.6
Taste and flavor	19.1	14.9	14.3	26.0
Cost	22.5	8.3	12.1	19.9
Preparation time	4.6	3.8	3.8	5.2
No answer	2.2	3.2	2.4	4.3
Total (%)	100.0	100.0	100.0	100.0
Test result (sample size)	***(1254)		***(1254)	

Check origin of food at time of purchase	Age group		Sex	
	<40 years	≥40 years	Females	Males
Always	60.6	66.3	68.7	44.2
Occasionally	30.1	29.6	26.9	42.4
Never	8.9	3.7	4.0	12.6
No answer	0.4	0.5	0.4	0.9
Total (%)	100.0	100.0	100.0	100.0
Test result (sample size)	***(1254)		***(1254)	

Having knowledge of food safety	Age group		Sex	
	<40 years	≥40 years	Females	Males
Yes	5.6	10.6	8.3	10.8
Some	29.9	47.9	41.7	39.0
Neither yes nor no	42.2	29.2	35.5	27.3
Little	18.2	10.0	12.3	16.0
No	3.7	2.0	1.9	6.1
No answer	0.4	0.4	0.3	0.9
Total (%)	100.0	100.0	100.0	100.0
Test result (sample size)	***(1254)		***(1254)	

Note: Significance level is the same as in Table 1.

These differences in self-awareness of metabolic syndrome were reflected in differences in exercise habits; those 40 years and older indicated more frequent exercise as well as greater health awareness.

Table 5 Awareness of health and environment issues

Seeing yourself as having metabolic syndrome	Age group		Sex	
	<40 years	≥40 years	Females	Males
Yes	5.0	13.3	8.2	19.1
Potentially yes	24.0	34.1	29.2	35.5
No	62.8	48.2	56.7	39.8
Not sure	7.6	4.2	5.5	5.2
No answer	0.7	0.3	0.4	0.4
Total (%)	100.0	100.0	100.0	100.0
Test result (sample size)	***(1254)		***(1254)	

Regular exercise habit	Age group		Sex	
	<40 years	≥40 years	Females	Males
Everyday	6.9	14.7	10.8	16.5
Several times a week	19.7	26.8	24.6	22.1
Once a week	16.5	15.2	15.4	16.9
Rarely	56.3	42.9	48.7	44.2
No answer	0.7	0.5	0.6	0.4
Total (%)	100.0	100.0	100.0	100.0
Test result (sample size)	***(1254)		+(1254)	

Change in volume of household trash	Age group		-	-
	<40 years	≥40 years	-	-
Increased	9.5	2.9	-	-
Somewhat increased	18.6	14.3	-	-
No change	54.8	45.3	-	-
Somewhat decreased	13.4	28.7	-	-
Decreased	3.0	8.1	-	-
No answer	0.7	0.8		
Total (%)	100.0	100.0	-	-
Test result (sample size)	***(1254)		-	-

Carries own shopping bag	Age group		Sex	
	<40 years	≥40 years	Females	Males
Always	18.4	29.9	28.9	11.3
Occasionally	36.4	35.6	37.2	29.9
Not often	16.5	16.2	15.4	19.9
Rarely	28.8	17.8	18.1	38.5
No answer	0.0	0.5	0.3	0.4
Total (%)	100.0	100.0	100.0	100.0
Test result (sample size)	***(1254)		***(1254)	

Note: Significance level is the same as in Table 1.

Although regular physical exercise was not common among respondents and only a minority selected daily exercise as a response, we observed significant differences with regard to sex and age group. More than half of those responders under the age of 40 rarely exercised while about 40% of those 40 years and older exercised once or several times a week. Although more male than female respondents exercised daily, the difference was just barely significant. Thus, those 40 years or older are definitely health conscious.

Table 6 Participation in community work and willingness to be a food education volunteer

Participation in community work	Age group		Sex	
	<40 years	≥ 40 years	Females	Males
Positively	0.9	4.2	2.4	5.2
Somewhat positively	10.0	16.7	14.7	12.1
Passively	26.0	30.3	29.6	24.7
No	63.0	47.5	52.4	56.7
No answer	0.2	1.4	0.9	1.3
Total (%)	100.0	100.0	100.0	100.0
Test result (sample size)		***(1254)		*(1254)
Experience & willingness to be a food education volunteer	**Age group**		**Sex**	
	<40 years	≥40 years	Females	Males
Having experience	0.9	3.4	2.3	3.5
Yes, if chances are available	56.7	59.3	61.1	46.3
No	41.8	31.9	33.1	46.3
No answer	0.7	5.3	3.5	3.9
Total (%)	100.0	100.0	100.0	100.0
Test result (sample size)	***(1254)		***(1254)	

Note: Significance level is the same as in Table 1.

As an aspect of environmental awareness, although most respondents answered that there was no change in the volume of household trash, a sharp contrast between age groups was observed; the portion of those indicating a decrease in volume was 36.8% of those aged 40 years and over while only 16.4% of those of under 40 years of age indicated a decrease. Conversely, the portion of those indicating an increase was 28.1% among respondents under 40 years of age but was only 17.2% in those 40 years of age or older. It is assumed that these differences came from the family life cycle and environmental awareness.

With regard to using one's own shopping bags instead of plastic bags provided by the store, there was no large difference between age groups in the category of only occasionally using personal shopping bags. There were, however, differences in the category of always using and rarely using such bags. Of those 40 years or older, 29.9% always used personal shopping bags whereas 28.9% of those of under 40 rarely used such bags. Further, nearly 40% of the males answered that they rarely carried their own bags. To summarize, environmental awareness was notably different between age groups and gender.

Local community work, and food education volunteering

Generally, participation in local community work was not frequent (Table 6). Even in the proactive 40 and over age group, only half participated and positive participation, including the "somewhat positively" response, was only 20.9%. Still, relatively speaking, those 40 and over were more actively involved than those under 40 (1% statistical significance).

As for experiences in participation as a food education volunteer and willingness to participate, whereas only a small percentage of respondents had such experiences, those 40 and over and the female respondents expressed a higher intention to participate than those under 40 and who were male (Table 6). In particular, nearly 50% of male respondents expressed no interest in this type of volunteer work. In short, although the middle-aged male respondents were interested in their own health issues, they were not enthusiastic about engaging in food education, probably because of being busy with daily duties.

The willingness to participate as a food education volunteer in relation to various areas of interest is shown in Table 7. The areas of interest differed from one age group to another; those 40 and over preferred cooking class for a healthy diet and also food safety issues while those under 40 selected a succession of food culture, agriculture, forestry and fishery experiences, and international exchange of food cultures. To summarize, it is safe to say that those 40 and over have higher awareness of food, health, the environment, and the local community than those under 40.

Estimation of determinant factor of the willingness to participate as a food education volunteer

Given the analytical model and results of the statistical test, we set up an estimation model.

$$V = \alpha_0 + \alpha_1 \, houshld_1 + \alpha_2 \, houshld_2 + \alpha_3 \, food_1 + \alpha_4 \, food_2 + \alpha_5 \, envrn + \alpha_6 \, health + \alpha_7 \, community + \mu \qquad (6)$$

Where, α_0 = constant, α_1 = parameters to be estimated, $(i = 1,..., 7)$
μ = stochastic error

Specifically, the explained variable that demonstrates the willingness to volunteer is based on binary data; unity is given if respondents have such an intention or have volunteer experience whereas zero is given if respondents have no intention. As for explanatory variables, two variables are used as household attributes; first, we considered the variable of sex because the opportunity cost for volunteer work is supposed to be lower in females than in males. Second, those households that have children 13 years of age or older (junior high school age and older) get unity while zero is given for those with children under 13 years old. This is because parents will have more leisure time due to a decrease in childcare work when children reach junior high school age. This variable also can be interpreted as a proxy variable representing the inter-generation differences at around the age of 40 years to avoid multicollinearity due to the correlation between the age of children and parents. Also there is a high correlation between the generation variable at 40 years old and other variables related to life stage. These household variables will reduce the opportunity cost, so that we can expect positive sign conditions.

We considered two variables for food awareness. The first variable is the frequency of celebrations involving serving traditional foods at home, which represents the degree of food awareness. The second is the degree of checking areas of origin of purchased food

Table 7 Areas of interest as a food education volunteer

Items Percentage of responses	Age group		Test results (sample size)	Sex		Test results (sample size)
	<40 years	≥40 years		Females	Males	
Cooking class for healthy diet	26.0	35.7	***(1254)	34.3	22.5	***(1254)
Improvement in eating habits	23.6	24.2	n(1254)	25.4	17.8	**(1254)
Food culture succession activity	23.6	19.8	+(1254)	22.4	16.0	**(1254)
Food safety and appropriate labeling	14.5	25.1	***(1254)	22.3	16.5	*(1254)
Food waste and recycling	13.9	12.1	n(1254)	12.6	13.4	n(1254)
Agriculture, forestry & fishery experiences	13.4	8.8	**(1254)	10.4	11.3	n(1254)
International exchange of food cultures	9.5	6.2	**(1254)	7.9	5.2	+(1254)

Note: Significance level is the same as in Table 1.

(always = 1, not always = 0). As the variable for environmental awareness, the degree of reduction in household trash emission (decreased = 1, not decreased = 0) was used. As for the variable of health awareness, we used the frequency of regular physical exercise (daily = 1, not daily = 0). These variables are expected to have positive signs. It is supposed from the statistical analysis above that those with positive participation in community work will be expected to be positive in participation in volunteer work as well (positive = 1, not positive = 0).

For estimation, we used a binary logit model because of the binary nature of the explained variable. Binary logit models have been quite widely used for evaluation of binary qualitative choices of behaviours due to their simplicity although this model is not capable of dealing with cases of multiple choices. The estimation results are shown in Table 8. Although we tried an ordered logit model by using a three-category explained variable (having experience of food education volunteer = 3, having willingness to volunteer = 2, no willingness = 0), the result was not good enough to adopt. Thus, we took the results of the binary logit model. As a reference, we showed the robust estimate of variance in addition to the standard estimate of variance. There was no distinctive difference between the standard and robust estimates in terms of parameters and significance levels.

Not every parameter was inconsistent with sign conditions, which means that these variables raise the willingness for volunteer activity by reducing the opportunity cost for that work. Among the estimated parameters, the community oriented attitude had the highest odds ratio, that is, 2.40, which means that this attitude is the most influential factor for willingness to volunteer. Then, being female, being middle aged, and having health awareness followed. Factors related to food awareness had positive effects on willingness to volunteer, but were not highly influential. The environmental factor was a barely influential factor because the parameter showed 20% significance, which is only used for reference. In short, those urbanites who are community oriented, female, middle aged, and having health awareness can comprise a potential volunteer group supportive of food education. The result here is not inconsistent with the previous study on volunteering as mentioned in the literature review (Bussell and Forbes 2001).

Conclusions

Food education aims at establishment of self-motivating behaviour leading to healthy dietary habits. This paper investigated the awareness of food education by a questionnaire survey mailed to 3,000 inhabitants in Matsudo and explored the factors determining the willingness to participate as a food education volunteer.

From conceptual considerations, we stipulated that educated food behaviour accompanied by volunteer behaviour exerts externalities to society by self-imposing internalization of the opportunity cost of conducting volunteer work, which set the empirical framework below.

First, from results of the questionnaire survey, the awareness of food education greatly differs with statistical significance between age groups and between males and females. Specifically, the age group under 40 years old has lower awareness and knowledge about food than does the older group. Although males are self-conscious regarding health issues to some extent, they are not highly conscious of food education. Thus,

Table 8 Estimation result of willingness to participate as a food education volunteer (binary logit model)

Factor	Explanatory variable	Parameter		Odds ratio
		Ordinary std. err.	Robust std. err.	
Respondent/household attribute 1	Respondent sex (female = 1, male = 0)	0.4409*** (2.79)	0.4409*** (2.84)	1.55
Respondent/household attribute 2	Having child ≥ 13 years (yes = 1, no = 0)	0.3546** (1.99)	0.3546** (2.03)	1.43
Food awareness 1	Frequency of celebrations at home with traditionalfoods	0.1713*** (3.31)	0.1713***(3.34)	1.19
Food awareness 2	Checking origin of food at time of purchase (always = 1, not always = 0)	0.2270*(1.77)	0.2270*(1.79)	1.25
Environment awareness	Change in emission of household garbage (decreased = 1, not decreased = 0)	0.2083+ (1.53)	0.2083 + (1.52)	1.23
Health awareness	Practice of regular exercise (daily = 1, not daily = 0)	0.3577*** (2.95)	0.3577*** (2.97)	1.43
Community awareness	Participation in community activities (positive = 1, not positive = 0)	0.8746*** (4.79)	0.8746*** (4.74)	2.40
-	Constant	−1.3451***(−4.80)	−1.3451***(−4.87)	-
Number of samples		1254		1.254
Log likelihood		−796.7120		−796.7124
LR Chi-square test		85.52***		73.75***

Note: Note: Significance level is the same as in Table 1. Figure in () is Z value.

females and those 40 years or older have positive attitudes toward food education and knowledge of food education.

Second, from the estimation result of the determinant function of willingness to participate as a food education volunteer, those inhabitants who had positive attitudes toward health and community issues expressed their willingness to volunteer. In short, our study results support the necessity and effectiveness of promoting food education to enhance not only food aspects, but also multiple aspects of daily life such as health and community awareness.

Consequently, it is necessary to enhance the awareness of food education among those under 40 years old who are busy raising children and working at jobs. On the other hand, the existence of generation gaps in terms of interest in and knowledge regarding food education suggests a complementary role between generations. For instance, those of the middle-aged generation and who are highly conscious of community issues can play a complementary role by supporting as food education volunteers the busy younger generation that is involved in raising children. Further research is necessary on effective measures for food education and evaluation of these educational results. Also, a civic-participatory food policy framework should be more closely scrutinized in the long run.

Competing interests
The author declares that there is no any competing interest among people and organizations concerned.

Authors' contributions

The first author mainly conducted a data analyses while the second author was mainly in charge of design of the survey in this study and the third author was mainly in charge of implementation of the survey. We shared the research results. All authors read and approved the final manuscript.

Acknowledgements

The questionnaire survey for this research was financed by the Municipality of Matsudo and the subsequent analysis was funded by Grants-in Aid for Scientific Research No. 24658191, Japan Society for the Promotion of Science (JSPS). The authors are grateful for the constructive comments from anonymous reviewers.

References

Adachi M (2008) Theories of nutrition education and promotion in Japan: enactment of the "Food Education Basic Law". Asia Pac J Clin Nutr 17(S1):180–184

Allison LD, Okun MA, Dutridge KS (2002) Assessing volunteer motives: a comparison of an open-ended probe and Likert rating scales. J Community Appl Soc Psychol 12:243–255

Arai K (1995) The Economics of Education: An Analysis of College-going Behavior. Springer, Tokyo, pp 15–16

Blaylock J, Smallwood D, Kassel K, Variyam J, Aldrich L (1999) Economics, food Choices, and nutrition. Food Pol 24:269–286

Bowman W (2009) The economic value of volunteers to nonprofit organizations. Nonprof Manag Leader 19(4):491–506

Brown E (1999) Assessing the value of volunteer activity. Nonprof Volunt Sec Q 28(3):3–17

Bussell H, Forbes D (2001) Understanding the volunteer market: the what, where, who and why of volunteering. Int J Nonprof Volunt Sec Market 7(3):244–257

Cabinet Office (2006) White Paper on Food Education (Syokuiku Hakusyo) 2006 (in Japanese). Jijigahosya, Tokyo

Cabinet Office (2007a) Result of Consciousness Survey Concerning Food Education (in Japanese). Cabinet Office, Tokyo

Cabinet Office (2007b) White Paper on Food Education (Syokuiku Hakusyo) 2007 (in Japanese). Jijigahosya, Tokyo

Callow M (2004) Identifying promotional appeals for targeting potential volunteers: an exploratory study on volunteering motives among retirees. Int J Nonprof Volunt Sec Market 9(3):261–274

Cash SB, Sunding DL, Zilberman D (2005) Fat taxes and thin subsidies: prices, diet, and health outcomes. Food Econ-Acta Agric Scand C 2:167–174

Chang HH, Nayga RM Jr (2011) Mother's nutritional label use and children's body weight. Food Pol 36:171–178

Chappell MJ, LaValle LA (2011) Food security and biodiversity: can we have both? an agroecological analysis. Agr Hum Val 28:3–26

DeLind LB (2011) Are local food and the local food movement taking us where we want to go? or are we hitching our wagons to the wrong stars? Agr Hum Val 28:273–283

Edwards JSA, Hartwell HH (2002) Fruit and vegetables-attitudes and knowledge of primary school children. J Hum Nutr Diet 15:365–374

Evenson RE (1981) Food policy and the new home economics. Food Pol 6:180–193

Freeman RB (1997) Working for nothing: the supply of volunteer labor. J Labor Econ 15(1):S140–S166

Griffin M, Sobal J, Lyson TA (2009) An analysis of a community food waste stream. Agr Hum Val 26:67–81

Hager M, Brudney JL (2005) Net benefits: weighing the challenges and benefits of volunteers. J Volunteer Admin 23 (1):26–31

Handy F, Srinivasan N (2004) Valuing volunteers: an economic evaluation of the net benefits of hospital volunteers. Nonprof Volunt Sec Q 33(1):28–54

Handy F, Cnaan RA, Brudney JL, Ascoli U, Meijs LCMP, Ranade S (2000) Public perception of "Who is a Volunteer": an examination of the net-cost approach from a cross-cultural perspective. Int J Volunt Nonprof Organ 11(1):45–65

Heslop LA, Madill J, Duxbury L, Dowdles M (2007) Doing what has to be done: strategies and orientations of married and single working mothers for food tasks. J Consum Behav 6:75–93

Hughner RS, McDnagh P, Prothero A, Shultz CJ II, Stanton J (2007) Who are organic food consumers? A compilation and review of why people purchase organic food. J Consum Behav 6:94–110

Iyer ES, Kashyap RK (2007) Consumer recycling: role of incentives, information, and social class. J Consum Behav 6:32–47

Kanagawa Prefecture (2007) Summarized Result of Food-consciousness Survey (in Japanese). Kanagawa Prefecture, Yokohama

Kemp K, Insch A, Holdsworth DK, Knight JG (2010) Food miles: do UK consumers actually care? Food Pol 35:504–513

Kenkel DS, Manning W (1999) Economic evaluation of nutrition policy or there's no such thing as a free lunch. Food Pol 24:145–162

Kerton S, Sinclair AJ (2010) Buying local organic food: a pathway to transformative learning. Agr Hum Val 27:401–413

Kimura AH (2011) Food education as food literacy: privatized and gendered food knowledge in contemporary Japan. Agr Hum Val 28:465–482

Lautenschlager L, Smith C (2007) Beliefs, knowledge, and values held by inner-city youth about gardening, nutrition, and cooking. Agr Hum Val 24:245–258

Maff (2010) What is "Shokuiku (Food Education)? Ministry of Agriculture, Forestry and Fisheries of Japan, Tokyo, www.maff.go.jp/e/pdf/shokuiku.pdf. Accessed 13 August 2010

Mah CL (2010) Shokuiku: governing food and public health in contemporary Japan. J Sociol 46(4):393–412

Matsudo City (2008) Basic Program for Food Education in Matsudo (in Japanese). Matsudo City, Matsudo

McGinnis JM, Meyers LD (1999) Public policy and healthy eating. Food Pol 24:335–341

McMichael P (2000) The power of food. Agr Hum Val 17:21–33

Mezirow J (1997) Transformative learning: theory to practice. New Dir Adult Cont Educ 74:5–12

Mezirow J (2003) Transformative learning as discourse. J Transform Educ 1:58–63

Mook L, Sousa J, Elgie S, Quarter J (2005) Accounting for the value of volunteer contributions. Nonprof Manag Leader 15(4):401–415

Ohe Y (2011) Evaluating internalization of multifunctionality by farm diversification: evidence from Educational Dairy Farms in Japan. J Environ Manag 92:886–891

Ohe Y (2012) Exploring the educational function in agriculture: Evidence from Japan. In: Caron RM (ed) Educational Policy in the Twenty-first Century. Nova Science Publishers, New York, pp 225–244

Pivarnik LF, Patnoad MS, Richard NL, Gable RK, Hirsch DW, Madaus J, Scarpati S, Carbone E (2009) Assessment of food safety knowledge of high school and transition teachers of special needs students. J Food Sci Educ 8:13–19

Robinson R, Smith C (2003) Associations between self-reported health conscious consumerism, body-mass index, and attitudes about sustainably produced foods. Agr Hum Val 20:177–187

Roe B, Teisl MF (2007) Genetically modified food labeling: the impacts of message and messenger on consumer perceptions of labels and products. Food Pol 32:49–66

Scott AR, Pope PE, Thompson BM (2009) Consumer's fresh produce food safety practices: outcomes of a fresh produce safety education program. J Food Science Educ 8:8–12

Twiss J, Kickinson J, Duma S, Kleinman T, Paulsen H, Rilveria L (2003) Community gardens: lessons learned from California healthy cities and communities. Ame J Public Health 93(9):1435–1438

Visschers VHM, Hartmann C, Leins-Hess R, Dohle S, Siegrist M (2013) A consumer segmentation of nutrition information use and its relation to food consumption behaviour. Food Pol 42:71–80

von Normann K (2009) The impact of lifestyles and food knowledge on the food patterns of German children. Int J Consum Stud 33:382–391

Walsh A, Nelson R (2010) The link between diet and health: an exploratory study of adolescents in Northern Ireland using foodmaps. Int J Consum Stud 34:190–195

9

The perfect storm of business venturing? The case of entomology-based venture creation

Stefano Pascucci[1*], Domenico Dentoni[1] and Dimitrios Mitsopoulos[2]

* Correspondence:
stefano.pascucci@wur.nl
[1]Management Studies Group,
Wageningen University,
Hollandseweg 1, 6700EW,
Wageningen (NL)
Full list of author information is
available at the end of the article

Abstract

In this paper we discuss how cooperation and trust among entrepreneurs can be challenged when they are dealing with venture creation in the context of radical innovation. Entomology-based foods are considered as one of the most promising innovation in the food sector. However they impose radical changes in food consumption habits with high risk of low consumer acceptance. Four European entrepreneurs joined forces in a new venture operating in this sector, trying to make it a successful business. We asked two of them to participate in a venture creation game experiment. The results indicate that high individualized pay-offs can lead the entrepreneurs to deviate from trust and cooperation.

Keywords: Venture creation; Radical innovation; Insects; Prisoner's Dilemma

Background

The global population is growing rapidly and it is likely to reach 9 billion by roughly the middle of this century (United Nations 2014). It is also predicted that more than 7.5 billion inhabitants will be in the less developed countries while the population of the least developed countries is projected to reach 1.7 billion (United Nations 2014). At the same time the world population is increasingly urbanizing. United Nations predictions indicate that about 70% of the population to be living in cities by 2050 (United Nations 2014). This rapid urbanization triggers a growing consumption of meat which for developing countries represents the most concentrated source of vitamins and minerals (Tilman et al. 2001). In addition to that, the growing wealth in developed countries as well as emerging economies, such as China, increases the purchase power of the consumers and thus pushing for a greater demand of processed food from meat, fish and dairy (Tilman et al. 2001). The market globalization triggers even more the consumption of meat products and decreases the types of consumed food products (Yen 2009). Moreover still an important percentage of world population do not have access to sufficient proteins (sometimes reaching extreme hunger levels), and even more people suffer from a form of micronutrient malnourishment (Barrett 2010).

Against this background, food and feed from insects appears a promising way to cope with the abovementioned issues (DeFoliart 1997). Insects nowadays are already a major or secondary source of protein elsewhere in the world (i.e. Asia, Africa, Australia and South America). In these areas insects have been a valuable and integral part of the human diet for hundreds of years (DeFoliart 1999) For many countries insects also significantly contribute to the local economies (DeFoliart 1999). In recent years attention for insects as new source of food ingredients has raised also in industrialised (Western) countries (Pascucci and de-Magistris 2013). Many concerns have been raised as well, due to regulatory, cultural and psychological barriers to create and develop an entomology-based food and feed industry (Derkzen et al. 2010). However, despite the scepticism of many stakeholders, several start-ups are trying to step in this industry, thus facing all kind of challenges from both a regulatory and marketing point of view. The co-existence of regulatory and marketing challenges makes the venture creation process a unique case, a kind of "perfect storm" of business venturing in the context of agri-food sector, thus calling for specific attention of both academics and practitioners.

In this paper we focus particularly on *"how to create a vivid, profit making entomology-based company"* in industrialized country context, rather than on general cultural and psychological issues related to insect consumption. In order to answer this research question we first start discussing the issue of new venture creation and radical innovation in the agri-food sector, since eating insects can be perceived as radically innovative in developed countries. We then analyse in more details, what are the main challenges when it comes to setting up a new venture dealing with a radically innovative product, particularly in terms of trust and cooperation within the entrepreneurial team. Based on this conceptual approach we have set-up a "venture creation game experiment" with two entrepreneurs of an European start-up company in the sector of entomology-based food products. In this way we have been able to highlight the main challenges and opportunities of such a business, and namely the role of trust and cooperation. We provide indications for other entrepreneurs and academic scholars who are dealing with venture creation and radical innovation, especially in the agri-food sector. While consumer acceptance remains the dominant (background) strategic issue for an entomology-based start-up company, our results indicate that trust and cooperation between business partners is one of the main challenges for the success of a start-up dealing with a radical innovation.

Methods

Literature review

Insects as sustainable source of protein

From an historical point of view considering insects as food is not something new in Western societies. Greeks and Romans were consuming several kinds of insects (like grasshoppers and beetle larvae) which they considered delicacies (DeFoliart 1999). Nowadays consumers eat ingredients derived from insects mainly as additives or accidentally through the food processing. For example, an additive called carmine or cochineal is commonly used to add a red/pink colour to many food products such as yogurts, candies and drinks (Greenhawt and Baldwin 2008). This ingredient is a pigment of a bright-red colour obtained from specific insects (*cochinidae*), and it is coded as E120 in the food labelling process. A couple of years ago, the US Food and Drug Administration set up a rule requiring food companies to list cochineal extract and carmine on their labels when they are used in food products or cosmetics, but it does not require companies to indicate that these ingredients are derived from insects (van Huis 2013).

In general the level of protein and fats in all insect species is high, and on average much higher than the traditional sources of proteins (Cerritos 2009). The nutritional characteristics of insect proteins are also very interesting. Some insects have proteins of high quality and of high digestibility (77–98%), and the concentration of essential amino-acids is also high (46–96% of the nutritional profile) (Verkerk et al. 2007). In addition to that, insects have many vitamins, provide high energy, minerals and fibre. Some insects have higher contents of calcium, magnesium, iron, and potassium than those of most food products of vegetable or animal origins. The caloric value of some insects is 50% higher than soybeans, 87% higher than maize, 63% higher than beef, and 70% higher than fish and beans (Verkerk et al. 2007).

Raring insects can also have many potential advantages in comparison with the current livestock production. Insects need fewer inputs to give production as they have high efficiency to convert biomass to protein (Oonincx and de Boer 2012). Of course the food conversion efficiency depends on high-quality diets, and livestock conditions (DeFoliart 1997; Cerritos 2009). Moreover, insects as poikilothermic animals do not spend large amounts of energy and nutrients to maintain constant their body temperatures. Thus, they are more efficient in transforming plant biomass into animal biomass (DeFoliart 1999). Insects are also big energy saver as they produce less waste in terms of manure and ammonia (Oonincx and de Boer 2012). In addition to that, insects can utilize many of the indigenous resources not used by humans, as well as organic wastes (Ramos-Elorduy 1997).

The first introduction of insects in the food supply chain was as an alternative for the expensive fish meal and for pet food industry as there was an increasing need of high-protein feed. However until now insects have not been massively produced for human consumption in Western societies (Pascucci and de-Magistris 2013). Using insects as source of ingredients for food products is a quite radical idea in this context, and it requires dramatic changes in the existing and dominating food consumption habits (Pascucci and de-Magistris 2013; Derkzen et al. 2010). This is probably the main reason why insect-based food products can be considered a radical innovation in the agri-food industry. It is not just a minor incremental extension of current food products, but a more a ground-breaking concept. For the mass production of this novel food there is a need for technology which is a bundle of brand new knowledge, skills and equipments (Carayannis et al. 2003). Often innovation in the agri-food industry combines technological innovation with social and cultural innovation (Earle 1997; Capitanio et al. 2009). Insect food is not just a novelty or an improvement, it is a fundamental change, a cultural "step-jump" in terms of food consumption habits. Often, radical innovations are not introduced on the market because consumers are very risk-adverse, and they are reluctant to accept new products. Radical product innovations are new to their users, or are radical in terms of creating disruptions in existing usage patterns (Heiskanen et al. 2007). This kind of innovations breaks with traditions in their field. There are many similarities between consumer acceptance of radical products and technologies and food innovations. However due to both cultural and psychological issues related to insect-based food products, risks of consumer rejection and failures is even higher than other "technology-based" radical innovations (Pascucci and de-Magistris 2013).

New venture creation in the context of radical innovation

The failure rate of new ventures is in the most optimistic research estimated to be around 46% (Timmons and Spinelli 2009 p.106). This number shows the huge difficulties most entrepreneurs face in creating their company. In this section an overview is provided of what is written about the entrepreneurial process and the steps in order to create a New Venture (NV).

New Venture Creation (NVC) starts with an idea. To verify if an idea is a valuable opportunity, entrepreneurs need to invest in thorough research and investigate organizational as well as market feasibility. In assessing the feasibility of a NV opportunity there are basically four primary areas entrepreneurs can look at (Gundry and Kickul 2007 p. 63–64): (i) team dynamics, such as cooperation and trust, (ii) availability of resources, (iii) knowledge and information and (iv) ability to generate revenue. Particularly understanding the quality of team dynamics requires time in order to find out which team members fit the entrepreneurial idea, whether they are good co-operators, and assess their trustworthiness (Timmons and Spinelli 2009 p. 188). Solid and cooperative teams are far more likely to attract venture capital (Mason and Stark 2004) than companies that do not. It is clearly stated that in order to create a vivid NV, there must be trust, cooperation and reciprocity between team members. Thus putting together key team members and create trust between them is a crucial part of NVC, because "who is added to a team affects not just the content or the capacity of the team, but also how the team does what it does" (Forbes et al. 2006). The leader of a NV has to show interest, create and spread passion and have a clear vision of where to go. However balancing individual and team/collective interests is a difficult task often leading a prisoner's dilemma condition in NV (Cable and Shane 1997). Cooperation among partners is also important to define the right financing strategy of the NV and to cover initial start-up costs. In the initial financing stage, most companies get financed by personal savings, loans from friends and relatives (with or without interest) (Bessant and Tidd 2007 p. 276), and/or the use of a personal credit card (Gundry and Kickul 2007 p.177). Setting up the right incentives and balancing between individual and team-based risks and pay-offs is a key-factor for successful NVC. Moreover because financial resources are often very scares for NV, an often used strategy is to align with complementary partners and form strategic alliances (Koza and Lewin 1998). This can lead to advantages such as product improvement, technology advancements, increase of future strategic planning capabilities. Also the process of creating strategic alliances, therefore expanding the original team to other members and groups, requires trust and cooperation among entrepreneurs in the NVC process.

While VC is challenging per se, associating it to creation of new products and/or processes can be even more challenging. Innovations follow a similar pattern as VC. They start with an idea, followed by the evaluation of that idea. A way to think about innovation is also to think about the degree of novelty it is bringing up. In this respect we have to make a distinction between an incremental innovation and a radical innovation, or "Doing what we do better" versus "New to the world" (Bessant and Tidd 2007 p.15). In order to examine whether a product is really new-to-the-world, Markides and Geroski (2005 p.4), posed two conditions which have to be met:

1) They offer new value propositions that radically change existing consumer habits and behaviour.
2) "The markets they create undermine the competences and complementary assets on which competitors built their success".

When NVC is associated to a radically innovative (RI) product is calling for understanding very specific circumstances. As distinct from NVC with a "normal" product, setting up a business with a RI product means taking higher risk and, in most cases, involving uncertainty of the highest degree. When it comes to innovation in NVC, we can consider different elements (Bessant and Tidd 2007 p. 40) such as personal or individual level which focuses on the role of creativity and entrepreneurship, the collective or social level that focuses on the importance of team dynamics and the context. Creativity involves not only finding bright ideas; it also deals with the thousands of problems arising in the NVC process. This creativity needs to be in balance with control on time, money and key resources. In order to establish a successful new venture with a RI product, entrepreneurs need to develop leadership and rely on several factors (Bessant and Tidd 2007 p.42) such as the ability to gather information from a wide range of sources, ability to give meaning to information, ability to reduce/change routines/existing behaviours and the ability to solve problems at an early stage. Entrepreneurial teams dealing with RI also need to rely on complementary skills, thus combining all kinds of knowledge to give the NV valuable insights on marketing aspects, financial issues and product development. Moreover they need to develop specific features of leadership. For example due to high uncertainty, the leader and the management team should have a clear overview of the process. They need to develop the ability to work together such that the structure of the organisation can allow people to deploy their creativity and share their knowledge. (Bessant and Tidd 2007 p.19 and p.27). It is also highlighted the relevance of learning from mistakes and to build proactive links between boundaries in and outside the organisation like in a "multi-player game" (Bessant and Tidd 2007 p. 20). In order to maintain an innovation-friendly environment, entrepreneurs have to encourage experimentation. It also might help to visit other companies and customers' organisations regularly, and get the management team inspired by new ideas and other routines (Gundry and Kickul 2007 p. 306). Another way to ban out uncertainty in the NVC process with a RI product, is research and development (R&D). This mainly focuses on the product design part. In normal NVC, the R&D phase usually covers the first one and a half years (Timmons and Spinelli 2009 p. 309). For NVC with a radical innovation this might be even longer. For the amount of investment needed, probably the same holds. Also, there is the risk of failure to get financial support to support and develop your ideas. Support from inside and outside the organisation is in fact necessary.

The case study: a venture creation dilemma in "Company X[a]"

Challenges imposed by RI products are all very important for testing trust and cooperation in an entrepreneurial team dealing with a NVC. To empirically test this issue we developed a prisoner's dilemma based experiment with entrepreneurs involved in a real start-up company engaged in entomology-based products.

Company X is a venture located in Western-Europe funded by four entrepreneurs in the autumn 2011. They all have different backgrounds, namely in the technical field (LCA analyst), management and marketing. Moreover, they have a world-class chef in their team.

The general vision of Company X can be summarized as follow: *"insect is a promising solution to meet the growing demand for animal proteins when traditional sources are reaching their environmental limits"*. As showed by Figure 1 company X believes that insects should not be eaten for fun, as rare delicacies sold at a luxury price. Instead, they strongly believe that insects are a serious opportunity to improve food

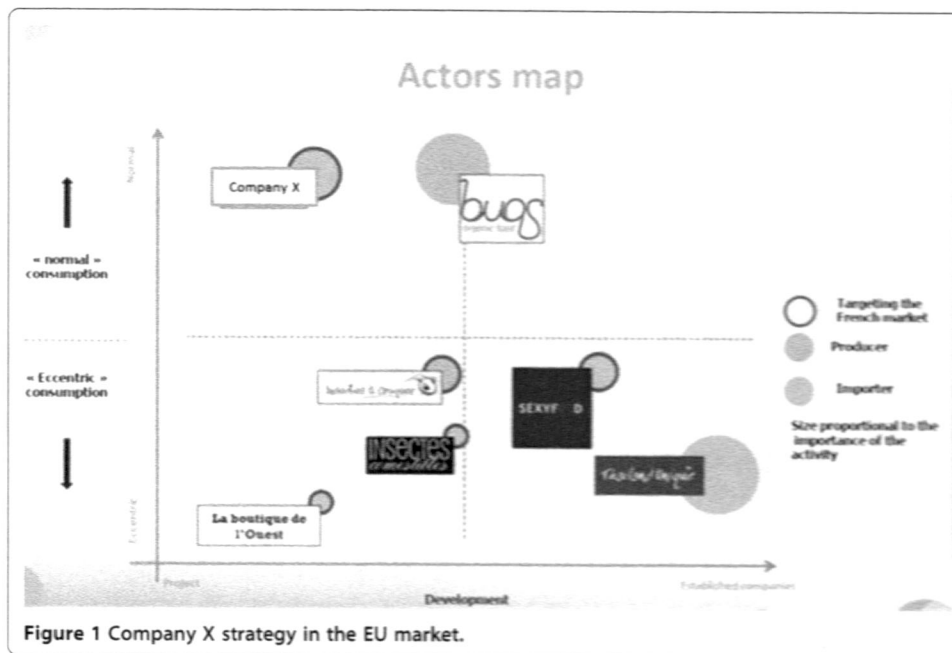

Figure 1 Company X strategy in the EU market.

consumption habits, because they show great value for health and the environment. This is differentiating their strategy compared to their competitors' ones (Figure 1). Coherent with their strategy the four entrepreneurs have settled a mission for achieving their mid-term objectives, such as "creating, preparing and sale on the European territory, products prepared with insects and intended for human consumption". Company X mid-term objective is to *produce and sell edible insect-based products, in which the insect will be made invisible*. Currently, the entrepreneurs are working on differential studies regarding biochemical properties of insects and consumer acceptance. The Management Studies (MST) department from Wageningen University (NL) has played an important role in the latter. Besides this, the company focuses on developing industrial-size processing plants. The cooperation between Company X and Wageningen University, aims at four marketing topics, as posed by Company X:

1. Analysing the mental representations regarding insects and more especially regarding insects as food;
2. Determining drivers and barriers to insect-based products;
3. Qualifying an quantifying the main targets;
4. Understanding the main targets of consumer habits in order to integrate our products in their frame of consumption.

The experiment was held in the late December 2011. To reduce hypothetical biases due to unrealistic options we inform the entrepreneurs that we were asking them to participate to a program lead by an European funding institution. We frame the participation indicating that the program would be funded in the next future conditional to budget availability. We framed the participation indicating that the results would have been considered for shortlisting potential candidates. We also asked to consider the participation in the program within the more general agreement Company X has arranged with Wageningen University. Two of the four entrepreneurs were asked to participate. They were the two in charge for financial and marketing strategy of the company respectively. We clarified the procedure with the two players and make sure they were understanding the role for the game.

More specifically we double-checked on the reward mechanism: only if both of them would have chosen the same solution in each option round, then they would have been considered as potential candidate to be really funded in a later stage. They both know that the probability of being successful would have been conditional on the other-player choice. We ran the experiment separately. We made sure that the two entrepreneurs couldn't communicate or transfer information.

The experiment has been designed following theoretical insights from Cable and Shane (1997). The essence of a Prisoner's Dilemma game is that "each individual actor has an incentive to act according to competitive, narrow self-interest even though all actors are collectively better off (receive higher rewards) if they cooperate" (Cable and Shane 1997 p. 145). To display this graphically, a general payoff matrix is introduced in Table 1:

Where R = reward when both players cooperate; P = punishment when both players defect; S = sucker's Payoff (penalty for cooperation while the other defects); T = temptation by extra payoff from defection (Cable and Shane 1997). Such that the payoffs for each player are dictated by the strategy adopted by the other player and follow the payoff structure $T > R > P > S$ (Cable and Shane 1997). In the context of this experiment[b] we decided to check for two elements of cooperation and trust:

1) whether the two players (entrepreneurs) were committed to the original strategy of the company (strategy A) or deviating for a strategy which was rewarding them more (strategy B);
2) whether the two players were committed to (lower) team-based rather than (higher) individual based rewards.

Therefore to let the experiment be incentive compatible and considering we were testing "temptation to defect", so whether the player is willing to give up "cooperation and trust" for an "individual-oriented reward", we made more attractive to deviate rather than cooperate. However punishing for simultaneous defection was not feasible. Therefore original payoff structure has been changed such that $T > R > P = S$. In other words we made the pay-off of being "sucked" and the punishment equal. Introducing a punishment payoff (such as a fine to be paid by one of the entrepreneur to the other one would have made the experimental setting unrealistic). However the incentive compatibility principle has been preserved.

In this way we could assess the value of cooperation against defection (individualised rewards). We used a change in strategy as a treatment to mitigate or increase

Table 1 Prisoners' dilemma game setting

		Player 2	
		Cooperates	Defects
Player 1	Cooperates	R,R	S,T
	Defects	T,S	P,P

"temptation to deviate". As said strategy A is reflecting the original strategy of Company X, which can be described as follow:

- Vision: "The insect is a promising solution to meet the growing demand for animal proteins when traditional sources are reaching their environmental limits"
- Mission: "The creation, preparation and sale on the European territory of products prepared with insects and intended for human consumption"
- Target consumers: "Normal consumers, willing to try new products but not driven by eccentric consumption habits. Avoiding niche and specialized markets. Supermarket as main outlet. Premium price for being healthy products and not luxury/exclusive products (such as delicatessen). Avoiding visualization of insects".

We proposed an alternative strategy that would possibly generate higher financial rewards and lower risks. We call it "strategy B". Since this strategy is challenging the current strategy, which was agreed in a team-based setting, we considered strategy B as an indication to deviate from cooperation and deviation from "trust in the group leadership". The vision and mission stay the same as in strategy A, but a different target group of consumers is addressed:

- Target consumers: "Environmental oriented and sensitive consumers, willing to experience new products. It is a niche product, targeting consumer buying groups or consumers cooperatives. Strong association with organic foods (premium price due to credence features). Avoiding visualization of insects".

The alternative business strategy is not yet proven to be a successful strategy for a new venture is the entomology-based business. Though it definitely has, according to preliminary marketing analysis, the potential to generate revenue with a relatively low amount of business risk (Table 2).

Where R = reward when both players cooperate; P = punishment when both players defect; S = sucker's payoff (penalty for cooperation while the other defects); T = temptation by extra payoff from defection (Cable and Shane 1997). Such that the payoffs for each player are dictated by the strategy adopted by the other player and follow the payoff structure $T > R > P = S$. The option choice as displayed in Table 3 is one of the 8 decision-scenarios used in the case of Campany X^c. In this example the two entrepreneurs have to choose between an individual-based reward (10,000 euro) and a team-based award (divide €10.000 over 4 team members). The outcome of such a venture creation game is that both entrepreneurs are better off when they cooperate, because in case of one entrepreneur

Table 2 Venture creation game for testing defection from cooperation

		Player 2	
		Cooperates (Team-based reward)	Defects (Individual-based reward)
Player 1	Cooperates (Team-based reward)	R,R	S,T
	Defects (Individual-based reward)	T,S	P,P

Table 3 Example of a pay-off matrix in one of the option of the venture creation game

| | | Player 2 | |
		Cooperates	Defects
Player 1	Cooperates	2500, 2500	0, 5000
	Defects	5000, 0	0,0

defecting, they do not get anything. This makes clear that cooperation in an entrepreneurial management team is very important.

Results and discussion

In Table 4 we present the results obtained in the experiment. As said the entrepreneurs were exposed to 8 decision-scenarios in which they have to decide whether to opt for a cooperative rather than individualistic solution. In the first two scenarios cooperation was achieved since both entrepreneurs indicate team-based stakes as preferred to individual based. Scenario 1 can be considered as the benchmark since it is proposing no incentive to deviate (both in terms of financial stakes and strategy seeking). In scenario 2 a small incentive to deviate was introduced, since strategy treatment would have lead for a less risky solution. However from scenario 3 to scenario 8 player 2 systematically deviated for more individualistic based rewards.

In an ex-post interview we asked the two entrepreneurs to motivate their decisions. While player 1 indicated that he was keen on pursuing team-based options, no matters the stakes indicated in the experiment, player 2 was indicating budget-seeking behaviour at the base of his choice. In short she indicated that given the early stage of their business, it was more important to look at options where more financial resources were provided. He also argued that at this stage it was not relevant whether financial resources would have been granted to the team or at least to one of its member. Player 1 is the initiator and main promoter of the venture, while player 2 is the expert in marketing and institutional relations. The different role they play in the company can explain their different decisions.

Conclusions

In recent years attention for novel sources of food ingredients and especially proteins is raising worldwide. Particularly in industrialised countries the idea of using insects is gaining consensus among a number of young start-ups. However despite their optimism, a number of concerns have been raised due to regulatory, cultural and

Table 4 Results of the venture creation game with Company X

Scenario	Strategy treatment	Individual-based reward (T)	Team-based reward (R)	Result
1	No	2,500	2,500	Cooperation
2	Yes	2,500	2,500	Cooperation
3	No	5,000	2,500	Player 2 defected
4	Yes	5,000	2,500	Player 2 defected
5	No	10,000	5,000	Player 2 defected
6	Yes	10,000	5,000	Player 2 defected
7	No	20,000	5,000	Player 2 defected
8	Yes	20,000	5,000	Player 2 defected

psychological barriers. In this paper we have framed the co-existence of these regulatory and marketing challenges as the "perfect storm" of business venturing in the agri-food sector. Particularly we have focused on the issue of *how to create a vivid, profit making entomology-based company*" in industrialized countries, rather than on general cultural and psychological issues related to insect consumption. We tried to answer this research question firstly by discussing the issue of new venture creation and radical innovation in the agri-food sector. Literature and empirical evidence highlight that new venture creation is a risky operation. When dealing with a radical (food) innovation the risks are even higher. There are three clear challenges which can be derived from the literature review such as issues of (i) consumer acceptance, (ii) financial resource capabilities, and (iii) building strategic alliances. However before dealing with those challenges, entrepreneurs involved in new venturing have to deal with ensuring a good level of cooperation and trust in their business environment. When dealing with radically innovative products this is even more important. Starting form this conceptual consideration, we have analysed in more details what are the specific challenges when it comes to setting up a new venture dealing with a radically innovative product, in an entomology-based start-up (company X). More specifically we have set-up a "venture creation game experiment" to highlight the main challenges and opportunities of such a business, and namely the role of trust and cooperation. Results indicate that while consumer acceptance remains the dominant strategic issue for an entomology-based start-up company, trust and cooperation between entrepreneurs is one of the main challenges for the success of a start-up dealing with a radical innovation. Therefore a vivid, profit-making entomology-based company is best created when the entrepreneurial team has been based on strong trust and cooperation principles. What has to be kept in mind is that total commitment to the team is essential and the prisoner's dilemma based experiment has been a way to find out the team cohesion and willingness to cooperate.

Of course given the number of observations and the specificity of this type of food innovation we are carefully constrained in drawing more general conclusions. However the use of experiments to assess entrepreneurs behaviour, their "social preferences", such as trust and cooperation, is indeed encouraging. The next step will be to implement experiments at a larger scale and to use as real as possible pay-off. This will increase the realism of results and the power of prediction of the analysis.

Endnotes

[a]We use "Company X" to preserve confidentiality. More detailed information on the company and the experiment can be asked to the authors.

[b]The description of the experiment is reported in Appendix A

[c]See Appendix A for all the other options proposed in the experiment.

Competing interests
The authors declare that they have no competing interests.

Authors' contributions
SP prepared the background section, the case study analysis section, the results and discussion section. DD the conclusions section. DM the literature review. All authors read and approved the final manuscript.

Author details
[1]Management Studies Group, Wageningen University, Hollandseweg 1, 6700EW, Wageningen (NL). [2]MSc Organic Agriculture, Consultant in Agribusiness Marketing and Innovation, Wageningen (NL).

References

Barrett CB (2010) Measuring food insecurity. Science 327(5967):825–828

Bessant J, Tidd J (2007) Innovation and entrepreneurship. John Wiley & Sons Inc, Chichester, England, p 462

Cable DM, Shane S (1997) A prisoner's dilemma approach to entrepreneur-venture capitalist relationships. Academy of Management Review 22(1):142–176

Capitanio F, Coppola A, Pascucci S (2009) Indications for drivers of innovation in the food sector. British Food Journal 111(8):820–838

Carayannis EG, Gonzalez E, Wetter J (2003) The nature and dynamics of discontinuous and disruptive innovations from a learning and knowledge management perspective, vol The international handbook on innovation (Part II)

Cerritos R (2009) Insects as food: an ecological, social and economical approach. CAB Rev: Perspect Agr Vet Sci Nutr Nat Resour 4(027):1–10

DeFoliart GR (1997) An overview of the role of edible insects in preserving biodiversity. Ecol Food Nutr 36(2–4):109–132

DeFoliart GR (1999) Insects as food: why the western attitude is important. Annu Rev Entomol 44:21–50

Derkzen P, Dries L, Pascucci S (2010) Conceptualizing the acceptance of insect-based food ingredients in western diets. Proceedings of the 5th International Consumer Sciences Research Conference on "Consumer Behaviour for a Sustainable Future". University of Bonn, Bonn, Germany

Earle MD (1997) Innovation in the food industry. Trends Food Sci Technol 8(5):166–175

Forbes DP, Borchert PS, Zellmer-Bruhn ME, Sapienza HJ (2006) Entrepreneurial Team Formation: An Exploration of New Member Addition. Entrepren Theor Pract 30(2):225–248

Greenhawt M, Baldwin JL (2008) Food colorings and flavors. Food Allergy: Adverse Reactions to Foods and Food Additives, 4th edn. John Wiley & Sons, New York, pp 403–428

Gundry LK, Kickul JR (2007) Entrepreneurship strategy – Changing Patterns in New Venture Creation, Growth, and Reinvention. Sage Publications, California, p 401

Heiskanen E, Hyvönen K, Niva M, Pantzar M, Timonen P, Varjonen J (2007) User involvement in radical innovation: are consumers conservative? Eur J Innovat Manag 10(4):489–509

Koza MP, Lewin AY (1998) The Co-Evolution of Strategic Alliances. Organ Sci 9(3):255–264

Markides C, Geroski PA (2005) Fast Second – How smart companies bypass radical innovation to enter and dominate new markets. John Wiley and Sons, San Francisco

Mason C, Stark M (2004) What do Investors Look for in a Business Plan? Int Small Bus J 22(3):227–248

Oonincx DG, de Boer IJ (2012) Environmental impact of the production of mealworms as a protein source for humans–a life cycle assessment. PLoS One 7(12):e51145

Pascucci S, de-Magistris T (2013) Information bias condemning radical food innovators? The case of insect-based products in the Netherlands. Int Food Agribusiness Manag Rev 16(3):1–16

Ramos-Elorduy J (1997) Insects: A sustainable source of food?. Ecology of food and nutrition 36(2–4):247–276

Tilman D, Fargione J, Wolff B, D'Antonio C, Dobson A, Howarth R, Schindler D, Schlesinger WH, Simberloff D, Swackhamer D (2001) Forecasting agriculturally driven global environmental change. Science 292(5515):281–284

Timmons JA, Spinelli S (2009) New Venture Creation – Entrepreneurship for the 21st Century (8th edition). McGraw-Hill, New York, p 666

United Nations (2014) World Urbanization Prospects: The 2012 Revision. Available at visit http://esa.un.org/wpp/ (Accessed at 17 July 2014)

van Huis A (2013) Potential of insects as food and feed in assuring food security. Annu Rev Entomol 58:563–583

Verkerk MC, Tramper J, Van Trijp JCM, Martens DE (2007) Insect cells for human food. Biotechnol Adv 25(2):198–202

Yen AL (2009) Edible insects: Traditional knowledge or western phobia? Entomologic Res 39:289–298

10

The adoption of technologies, management practices, and production systems in U.S. milk production

Jeffrey Gillespie[1*], Richard Nehring[2] and Isaac Sitienei[3]

* Correspondence: jmgille@lsu.edu
[1]Department of Agricultural Economics and Agribusiness, Louisiana State University Agricultural Center, Martin D Woodin Hall, Baton Rouge, LA 70803, USA
Full list of author information is available at the end of the article

Abstract

Adoption rates of 19 dairy technologies, management practices, and production systems (TMPPS) are estimated for the U.S. for 2005 and 2010 and, in cases where data are available, 1993 and 2000. Logit models are estimated to determine types of farms most likely to use each TMPPS. TMPPS experiencing the greatest increases in adoption have been automatic take-offs, the internet, breeding technologies, and USDA certified organic production; recombinant bovine somatotropin experienced a reduction in usage between 2005 and 2010. Factors influencing TMPPS usage include farm size, tenure, and diversification; farmer age and education; and region of the U.S. where the farm is located.

Keywords: Technology adoption; Breeding technology; Computerized technology; Production system

Background

Over the past two decades, the U.S. dairy industry has experienced significant structural change that has been accompanied by increases in the adoption of productivity-influencing technologies, management practices, and production systems (TMPPS). A number of factors can be attributed to changes in TMPPS usage by dairy farmers, including the capacity of some TMPPS to allow for realization of benefits associated with economies of size and/or improved efficiency, as well as changing consumer tastes and preferences for milk produced under specific production systems. The research presented in this paper adds to past literature on dairy technology adoption by analyzing the drivers that have influenced the use of TMPPS in the U.S. dairy industry, estimating new 2010 aggregate adoption estimates for these TMPPS, and analyzing the adoption diffusion of these TMPPS over the past two decades.

There have been large changes in numbers of dairy farms, total milk production, and milk production per cow in the U.S. over the past 20-years. From 1991 to 2010, total milk production increased by 30%, the number of farms decreased by 66%, and milk production on a per-cow basis increased by 42%, showing significant changes in the industry structure and productivity over the period. Clearly, average farm size increased along with cow productivity. A major contributor to this structural change has been the adoption of TMPPS that have allowed for greater economies of size and increased cow productivity.

Farmers generally adopt a TMPPS if it serves to increase farm profit and fits within the farm's resource constraints. In deciding whether to adopt a TMPPS, the farmer must consider whether it would change output quantity, output price, and/or cost of production. For example, use of recombinant bovine somatotropin (rbST) serves to increase milk produced per cow, but in areas where non-rbST premium milk prices are provided, a lower effective price per unit is received for milk produced under the technology. Furthermore, adoption would serve to alter production costs. In addition to the costs and returns associated with a TMPPS, farmers must consider whether adoption can occur given land, labor, capital, and credit constraints. For example, adoption of a pasture-based system may not fit within the farm's resource constraints if insufficient land is available for grazing or the land is in a higher-value use.

A number of studies have addressed factors influencing the adoption of various combinations of TMPPS in dairy production, most using limited dependent variable models to assess factors influencing usage. Table 1 describes each TMPPS and provides a brief summary of previous work for each. The most recent comprehensive study addressing TMPPS usage in U.S. dairy production was Khanal et al. (2010). The present study differs from theirs in several important ways. First, they analyzed 11 TMPPS; we analyze a fuller set, 19. Second, their analysis used mean comparison tests to compare usage on the basis of five farm and farmer demographic drivers, while we use multivariate analysis (logit) to analyze usage with 16 drivers, including farm, farmer demographic, and regional variables. Thus, we were able to fully account for the influence of adoption drivers used in the logit regression models. Third, their analysis focused on adoption drivers in 2000 and 2005 while ours focuses on the more recent 2010 data. Fourth, our analysis uses the full set of USDA Agricultural Resource Management Survey (ARMS) and Farm Costs and Returns Survey (FCRS) data, dairy versions, since 1993 to examine aggregate adoption over a 17-year period. Though this period is not sufficient to examine full diffusion of most technologies from introduction to equilibrium, it provides insight into TMPPS use patterns over an extensive period of diffusion.

The analysis of technology adoption and its determinants in agriculture has a rich history. In an early classic analysis of the technology adoption process, Griliches (1957) examined the diffusion of hybrid corn in the U.S., showing that technology diffusion followed an S-shaped, logistic curve from introduction to equilibrium. This shape implies that adoption diffusion starts off rather slowly, speeds up, and eventually levels out. He further showed strong influences of region on adoption rates, suggesting that identical adoption diffusion curves do not exist under all conditions. Cochrane (1958) discussed the agricultural treadmill, addressing how early technology adopters generally benefit the most economically, with late adopters being forced to either adopt or exit production. The analysis of technology adoption in agriculture increased throughout the 1960s and 1970s. By 1985, Feder et al. (1985) had reviewed the extensive literature on technology adoption in developing countries, discussing the major adoption drivers. The literature addressing technology adoption in agriculture has expanded since then, with a large body of work conducted on adoption determinants, the extent of which will not be fully discussed here. A substantial amount of this work has dealt with the U.S. dairy industry, much of which is discussed within this paper with respect to each of the TMPPS.

Table 1 Technologies, management practices, and production systems analyzed in the study

TMPPS	Description
Computerized and/or Automated Technologies	
Computerized feed delivery system	Provides specific feed ration to an individual or group of cows, depending upon cow's lactation phase. Typically used with total mixed ration designed to meet the animal's full nutritional needs.
Computerized milking system	At least compiles computerized milking data from milker, but may also refer to an automatic milking system or fully automated robotic system. Data provided useful for making individual cow decisions (Gillespie et al., 2009a).
On-farm computer to manage Dairy Records	Provides not only convenient way for farmers to keep farm records, but also facilitates analysis of the records.
Accessing the internet for dairy Information	May yield information on prices, input and output markets, and other useful data. Larger dairy farmers more prone to adopt technology and hold off-farm jobs have had greater internet experience (Grisham and Gillespie, 2008).
Automatic take-offs for milking units	Ensure cows are not under- or over-milked (resulting in increased mastitis incidence). Senses end of milk flow, shutting off milking unit vacuum and releasing it from the cow. Results in increased labor productivity and comfort.
Holding pen with an udder Washer	Facilitates automatic washing of cow teats prior to cow entering the milking parlor.
Breeding and/or Biological Technologies	
Artificial insemination	Involves artificially introducing semen into cow after collection from bull and processing. Introduced in the 1940s. Realized quick diffusion as a means to introduce superior genetics and eliminate transfer of venereal diseases (Foote, 1996), increase economic advantage (Barber, 1983), and avoid the need for farmers to deal with bulls. Khanal and Gillespie (2013) found higher profitability and lower costs with artificial insemination.
Sexed semen	Once collected from the bull, sperm separated into female X-bearing and male Y-bearing sperm cells prior to artificially inseminating. Advantage is ability to produce calves of desired sex (DeVries et al., 2008), though lower conception rates have been of concern (Wiegel, 2004).
Embryo transfer	Involves flushing embryos from a donor cow and transferring them to a recipient cow that is usually less valuable.
Recombinant bovine somatotropin	Released for commercial use in 1994 and expected to increase milk yield by about 10 lbs/day. Adoption and effect on farm profitability extensively studied (e.g., McBride et al., 2004; Tauer, 2009; Gillespie et al., 2010), with most finding no significant impact of use on farm cost and/or profitability. Modest adoption, partly due to negative consumer reactions to its use.
Management Practices	
Regular veterinary services	Promotes improved herd health and feed efficiency, with the objective of increasing production efficiency.
Use of a nutritionist to design feed mixes or purchase feed	Improves herd health and efficiency, with objective of increasing production efficiency. May help in curbing excretion of specific nutrients that are of environmental concern.
Keeping individual cow Production Records	Provides information on performance of each animal, assisting farmer in breeding, culling, and other decisions to increase production efficiency.
Forward purchasing of inputs	Involves contracting with an input supplier prior to purchase to ensure steady supply of inputs at a specified price, reducing risk.
Negotiating price discounts for dairy inputs	Facilitates procurement of lower-cost inputs. Generally more useful for larger operations that can purchase inputs in bulk. 32% of U.S. farms used forward contracts in 1994, with cash crop farmers using them more extensively than dairy farmers (Mishra and Perry, 1999).
Production systems	
Milk cows three or more times per day	Facilitates efficient parlor usage and increased cow productivity. Third milking increases milk production per cow 6% to 19% (Amos et al., 1985; DePeters et al., 1985). Additional yield similar across cows regardless of cow productivity (Erdman and Varner, 1995). Reduced reproductive

Table 1 Technologies, management practices, and production systems analyzed in the study *(Continued)*

	efficiency may result as cow spends more time being milked (Gisi et al., 1986).
Use of a dairy parlor	Compared to stanchion milking systems, use of a parlor generally reduces milking labor costs and is more cost-efficient for larger herds (Tauer 1998). Various configurations include swing, herringbone, parallel, side opening, polygon, carousel, flat barn. Cows enter stalls for milking, usually on raised platforms, and are released after milking.
Pasture-based milk production	Refers to extensive use of pasture for cow's forage needs. Provision of ≥50% of the cow's forage ration from pasture during the grazing months (Gillespie et al., 2009b). Generally produces less milk per cow, but at lower cost per cow. Some consumers willing to pay premium prices for milk from pasture-based operations. No significant differences found in profitability between pasture-based and similar-scale conventional milk production (Gillespie et al., 2009b).
USDA certified organic milk production	Has increased over past decade in response to consumer demand. Requires use of organically-grown feed and no growth promotants or antibiotics. U.S. organic milk production costs shown to be $5 - $8/cwt higher than for conventional milk (McBride and Greene, 2009). No significant differences in technical efficiency found between U.S. organic and non-organic dairy farms (Mayen et al., 2010).
Controlling the breeding and/or calving season	Refers to practice of synchronizing breeding, calving, lactation, and dry seasons in dairy herd. Most often used in pasture-based operations to exploit optimal pasture conditions throughout year, but also other advantages (Turner and Skele, 2007).

In this paper, in accordance with Feder et al. (1985), we use the term, "aggregate adoption" to refer to the portion of U.S. producers using each TMPPS. Adoption diffusion, on the other hand, refers to the "process of spread of a new (TMPPS)" (Feder et al., 1985), in our case throughout the U.S.

Technologies, management practices, and production systems in the U.S. dairy production

We categorize the major TMPPS used in U.S. dairy production as computerized and/or automated technologies, breeding and/or biological technologies, management practices, and production systems. Each TMPPS analyzed in this study is described and discussed in Table 1. Computerized and/or automated technologies refer to technologies that utilize computer hardware and software to enhance the efficient use of resources and/or provide information to the farmer. These technologies included in our analysis are *Computerized Feed Delivery System, Computerized Milking System, On-farm Computer to Manage Dairy Records, Accessing the Internet for Dairy Information, Automatic Take-offs for Milking Units,* and *Holding Pen with an Udder Washer.* Breeding and/or biological technologies refer to biological advances that result in greater reproductive and/or production efficiency. These technologies included in our analysis are *Artificial Insemination, Sexed Semen, Embryo Transfer,* and *rbST.*

Management practices refer to methods farmers use to impact productivity, with or without the use of a specific technology. Management practices included in our analysis are *Regularly Scheduled Veterinary Services, Use of a Nutritionist to Design Feed Mixes or Purchase Feed, Keeping Individual Cow Production Records, Forward Purchasing of Inputs,* and *Negotiating Price Discounts for Dairy Inputs.* Production systems differ if switching from one to another involves a fundamental change in the way the farm is managed on a daily basis. Production systems included in our analysis are *Milk Cows Three or More Times per Day, Use of a Dairy Parlor, Pasture-based Milk Production, Organic Milk Production,* and *Controlling the Breeding and/or Calving Season.* Note that use of our selected TMPPS does not necessarily increase production or efficiency. For example, pasture-based systems generally require more land and less variable input per unit of milk relative to conventional systems, while producing less milk per cow. Pruitt et al. (2012) similarly categorized TMPPS for U.S. cow-calf farms, without separating the technologies into subcategories.

Methods

Similar to the exposition provided by Pruitt et al. (2012), economic theory suggests the farmer maximizes expected utility associated with the adoption of TMPPS as:

$$\max_{i} EU(\pi|m) \tag{1}$$

where $i \in \{0,1\}$ with 0 indicating non-adoption and 1 indicating adoption. The $EU(.)$ operator indicates expected utility; π is profit, where $\pi = R_i - C_i$ (revenue less cost); and m are farm and farmer characteristics that impact adoption.

The logit model, which assumes a logistic distribution, is a limited dependent variable model that is appropriate for analyzing decisions where there is a yes/no (1–0) outcome, such as whether farmers adopt a TMPPS. Using the logit model (Greene, 2000, p.814).

$$Prob\ (TMPPS\ =\ 1|x)\ =\ \frac{e^{x'\beta}}{1+e^{x'\beta}} = \Lambda(x'\beta) \tag{2}$$

Parameters β reflect the impacts of changes in x on the probability of adopting the *TMPPS* and $\Lambda(.)$ indicates the logistic cumulative distribution function. Two assumptions in a logit analysis examining TMPPS adoption are that producers are either adopters or non-adopters and that a TMPPS is well-defined. In actuality, in some cases producers may be partial-adopters, for example using rbST on a subset of the animals on the farm. We do not have information on percentages of production on a farm covered by each TMPPS, so our analysis addresses whether the TMPPS was adopted for any portion of the farm's production. For the second assumption, it is recognized that for some TMPPS, i.e. a computerized milking system, a range of technologies may fit under that particular category. Due to data limitations, we cannot further subdivide the category, but analyze whether the producer adopted any TMPPS that falls under that category.

Using the logit model, β parameters cannot be directly interpreted other than for sign, thus creating the need for measures that can be used to explain the magnitude of an independent variable's influence over adoption. The marginal effect for a continuous variable using the logit model is calculated as in Greene (2000, p.816):

$$\frac{\partial E[y|x]}{\partial x}\ =\ \Lambda(\beta'x)\ [1-\Lambda(\beta'x)]\beta \tag{3}$$

Marginal effects for dummy variables are calculated as shown in Greene (2000, p. 817). The McFadden R^2 is used as an indicator of goodness of fit for the models. Independent variables included in the models consist of farm structure, demographic, and regional variables, as follow.

Independent variables

Farm structure variables

Farm size is measured as the number of cows in the operation, *Cows*, and the number of *Acres* on the farm. *Cows* is the average number of milk cows on the operation during the year, divided by 100 for ease of interpretation of marginal effects. As the number of cows increases, it is expected that adoption of TMPPS with associated size economies, such as computerized feed delivery systems, will increase. TMPPS that have been considered to be scale-neutral, such as rbST, have also been found to be more extensively used by larger-scale producers (McBride et al., 2004). *Acres* is a farm size measure that considers the extent of the land base. *Acres* is defined as total acres operated, including both owned and leased land, less farmland leased to others, divided by 100 for ease of interpretation of marginal effects.

The portion of farmland owned by the operator, *Owned*, allows for consideration of the impacts of land tenancy on TMPPS usage. Feder et al. (1985) discuss differential impacts of land tenancy on TMPPS usage, depending on the nature of the TMPPS. TMPPS that require substantial labor and investments in real estate assets (i.e., building improvements) would be expected to have lower usage by renters. *Specialization*, (Value of Production from Dairy) ÷ (Value of Total Farm Production), is expected to positively impact the use of TMPPS that reduce risk and/or require greater management. Greater specialization (lower diversification) generally exposes producers to more risk, increasing the attractiveness of TMPPS such as forward pricing.

Variables Organic Milk Production and Pasture-based Milk Production Systems differ in resource usage and, in some cases, output than their "conventional" counterparts, thus their inclusion as independent variables. The expected impact of these systems on adoption varies by TMPPS. For instance, since rbST is prohibited for USDA certified organic dairies and the marginal value product of its use would unlikely exceed the marginal factor cost for pasture-based production, expected signs for these systems would be negative for rbST. Observation suggests that USDA certified organic and pasture-based systems tend to be less technology-intensive in general. On the other hand, use of controlled breeding seasons would be expected to increase with either of these systems and good record-keeping is required for USDA certified organic production.

Farmer demographic variables

Two farmer demographic variables are included, farmer *Age* (the farmer's age divided by 10 for ease of interpretation of marginal effects) and whether the farmer holds a four-year *College* degree. Older farmers have generally been lower users of TMPPS for which they are unlikely to realize a full stream of benefits prior to retirement, or which change their management requirements, such as rbST (McBride et al., 2004). Farmers with more education are expected to be greater users of advanced TMPPS, as found by McBride et al. (2004) and Ward et al. (2008).

Regions

Nine U.S. production regions are included: *West* (Arizona-AZ, Colorado-CO, Idaho-ID, New Mexico-NM), *Pacific* (California-CA, Oregon-OR, Washington-WA), *Southeast* (Florida-FL, Georgia-GA), *Corn Belt* (Illinois-IL, Indiana-IN, Iowa-IA, Missouri-MO,

Ohio-OH), *Northern Plains* (Kansas-KS), *Appalachia* (Kentucky-KY, Tennessee-TN, Virginia-VA), *Northeast* (Maine-ME, New York-NY, Pennsylvania-PA, Vermont-VT), *Lake States* (Michigan-MI, Minnesota-MN, Wisconsin-WI), and *Southern Plains* (Texas-TX). These regions differ in dairy industry structure, with the *Northeast* and *Lake States* traditionally being the largest dairy regions and the *West* and *Pacific* regions experiencing more recent growth. The base region in our models is the *Lake States*.

Comparing 2005 and 2010 aggregate adoption rates

Aggregate adoption rates of 19 TMPPS for 2005 and 2010 are compared using pair-wise 2-tailed *t*-tests utilizing the delete-a-group jackknife estimation procedure. The delete-a-group jackknife estimation procedure is used because the ARMS data are collected via a complex survey design with both an area and list frame, rather than a model-based random sample most commonly used for statistical analysis. Dubman (2000) explains how to estimate *t*-tests utilizing this procedure with ARMS data. These tests allow for determination of whether there were significant differences in aggregate adoption rates in 2005 versus 2010 and, thus, whether significant adoption diffusion occurred over the 5-year period for each TMPPS. For 12 of the TMPPS, Khanal et al. (2010) conducted similar analysis for 2000 and 2005. We included their aggregate adoption rates in our graphical analysis in order to present an extended perspective of diffusion. Additional aggregate adoption estimates are made for 1993 and 2000 in cases where data are available, allowing for further investigation of diffusion (to be discussed in greater detail in the next section). In addition to reporting percentages of farms using TMPPS, we also examine percentages of the total milk quantity produced by farmers using each of the TMPPS in 2005 and 2010.

Data

Data for analyses conducted in this study are primarily from the 2005 and 2010 ARMS, dairy version, conducted by the USDA National Agricultural Statistics Service (NASS) and Economic Research Service (ERS). The ARMS is conducted annually to collect economic information on U.S. farms. Every year, enterprises are selected for more in-depth surveying so that enterprise cost of production, input usage, and production systems can be estimated. The dairy enterprise was surveyed in detail in 1993 (FCRS), 2000, 2005, and 2010 (ARMS). Logit TMPPS adoption models used only 2010 data while aggregate adoption rates were estimated using all available years. We tried pooling 2005 and 2010 data for the logit analyses in a manner similar to Gillespie et al. (2010), but were not able to satisfy the likelihood ratio index poolability test as discussed in Pesaran et al. (1999). We found that interaction terms would be required for the discrete variable indicating year with all other independent variables. The implication was that independent variables had differential effects on the dependent variable, depending upon year. This resulted in multicollinearity, convergence problems in some cases, and minimal additional insight, leading us to use only 2010 data for the logit analyses. To ensure that only commercial operations were surveyed, the operation must have milked at least 10 cows at any time during the year. States covered include those listed above in the *Regions* subsection; KS and CO were surveyed in 2010 only. The

ARMS collects information on costs and returns at both whole-farm and dairy enterprise levels, TMPPS usage, and farm and household characteristics.

Sample farms for ARMS are selected from a list maintained by NASS. Using stratified sampling, each farm represents other "like" farms in the population. The dataset contains expansion factors (weights) to allow for extrapolation to the U.S. dairy population of the states where the survey was conducted. These states represent 90% of the U.S. dairy farm population of farms with 10 or more cows. Data were collected consistently in both 2005 and 2010 (hand enumeration using a complex sampling scheme and broad national coverage), so results from both years can be compared. The 2005 (2010) data include 1,814 (1,915) observations, respectively. For several TMPPS, adoption estimates may be made from the 2000 ARMS dairy survey and the 1993 Farm Costs and Returns Survey (FCRS), both of which were conducted using similar methods and, thus, are comparable to the 2005 and 2010 ARMS. The 2000 ARMS dairy survey included 870 observations from the same states as those in the 2005 ARMS with the exception of ME and OR. The 1993 dairy FCRS included 695 observations. These data do not link the same farms; each year's data is a separate cross-section representing the dairy farm population for that year. For several TMPPS, particularly those in the Computerized and/or Automated Technologies category, it is acknowledged that advances have been made over the past two decades; for example, up-to-date computerized milking systems may look different from those installed in 1993. Thus, interpretation of results for those TMPPS must be made with that realization.

Consistent with Dubman (2000), the delete-a-group jackknife procedure with 30 replicates is used for deriving statistical estimates. For more information as to why this procedure is used to estimate standard errors using the ARMS data derived from a complex survey design, see the report by the National Research Council, Panel to Review USDA's Agricultural Resource Management Survey (2008).

Results and discussion
Computerized and/or automated technologies
Aggregate adoption rates for computerized and/or automated technologies are shown in Figure 1. In 2010, computerized feed delivery systems were used by 8.7% of farmers compared with 7.1% in 2005, but the difference was not statistically significant (Table 2). This is compared with estimated usage rates of 8.1% in 2000 (Khanal et al., 2010) and 6.3% in 1993 (Short, 2000), showing rather tepid diffusion response over the 1993–2010 period. The percentage of milk produced by farms using computerized feed delivery systems increased from 28.2% in 2005 to 37.7% in 2010 (Table 2). For 2010, an additional 100 cows or 100 acres increased usage by 0.7 or 0.3 percentage points, respectively (Table 3). A farmer owning 100% versus 0% of his or her land decreased usage by 2.9 percentage points. Increasing the percentage of farm returns from dairy by 1% increased usage by 0.12 percentage points. Two of the regional variables were significant: Southeastern and Southern Plains producers were less likely to have adopted than Lake States farmers. Overall, while percentages of farmers using computerized feed delivery systems changed little over the past two decades, the percentage of milk produced by farms using this technology increased, explained by increasing numbers of larger-scale farms using it, as supported by our results. Though significance of regions

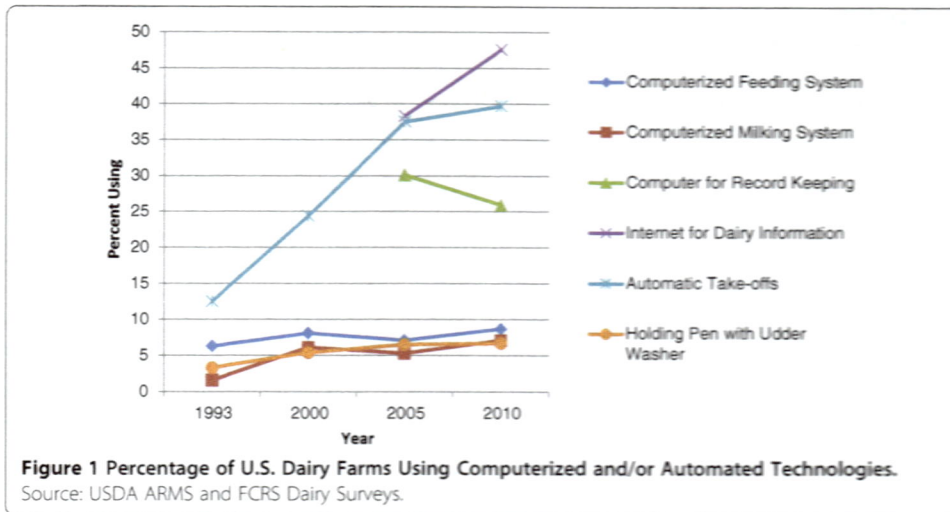

Figure 1 Percentage of U.S. Dairy Farms Using Computerized and/or Automated Technologies.
Source: USDA ARMS and FCRS Dairy Surveys.

is found for most TMPPS and shown in Table 3, in the interest of space, these results are not discussed further in the text.

The aggregate adoption rate for computerized milking systems increased from 5.3% in 2005 to 7.1% in 2010. These adoption rates are compared with the estimated aggregate adoption rates of 6.1% in 2000 (Khanal et al., 2010) and 1.6% in 1993 (Short, 2000), showing limited diffusion after 2000. The percentage of milk produced by farms using computerized milking systems was 22.8% and 24.2% in 2005 and 2010, respectively. Usage was greater on farms with more acreage, where a higher percentage of the land was owned by the farmer, and that were more heavily specialized in dairy. In 2010, holding a college degree increased usage by 5.6 percentage points. USDA certified organic farmers had usage rates that were 2.3 percentage points lower than those of non-organic farmers. Overall, computerized milking systems experienced modest diffusion over the past decade. Adoption drivers included specialization in dairy and the farmer's education level.

In 2010, the aggregate adoption rate of on-farm computers for record-keeping was 25.9% of farmers compared with 30.1% in 2005, but the difference was not statistically significant. The percentage of milk produced by farmers using an on-farm computer for record-keeping increased from 66.1% in 2005 to 73.1% in 2010. Larger-scale farmers and those holding college degrees were the greater users. Though use of on-farm computers for record-keeping does not appear to have increased from 2005 to 2010, use by larger-scale, more highly-educated producers producing greater percentages of the milk in 2010 suggests greater diffusion in the near future.

The percentage of farms using the internet for collecting dairy information increased from 38.2% in 2005 to 47.6% in 2010. The percentage of milk produced by farmers using the internet for dairy information increased from 63.6% in 2005 to 79.9% in 2010. In 2010, large-scale, specialized, and educated farmers were the greater users. Considering (1) larger farms run by more highly educated farmers are increasing in number, and (2) internet usage is generally diffusing throughout the general population, significant future diffusion is expected.

In 2010, the aggregate adoption rate for automatic take-offs for milking units was 39.7% of farmers, compared with 37.5% in 2005, but the difference was not statistically

Table 2 Adoption rates of technologies, management practices, and production systems (TMPPS) on U.S. dairy farms

Technology, Management Practice, or Production System	% of Farms Adopting 2005	% of Farms Adopting 2010	% of Milk Produced Covered by TMPPS 2005	% of Milk Produced Covered by TMPPS 2010
Computerized and/or Automated Technologies				
Computerized Feed Delivery System	7.1	8.7	28.2[c]	37.7[d]
Computerized Milking System	5.3[a]	7.1[b]	22.8	24.2
On-farm Computer for Records	25.9	30.1	66.1[c]	73.1[d]
Internet for Dairy Information	38.2[a]	47.6[b]	63.6[c]	79.9[d]
Automatic Take-offs for Milking Units	37.5	39.7	72.9	77.0
Holding Pen with an Udder Washer	6.5	6.7	31.4	30.7
Breeding and/or Biological Technologies				
Artificial Insemination	81.5	80.1	88.9[c]	92.2[d]
Embryo Transfer and/or Sexed Semen	10.4[a]	17.8[b]	15.7[c]	24.3[d]
Recombinant Bovine Somatotropin	16.6[a]	8.5[b]	40.0[c]	16.6[d]
Management Practices				
Regularly Scheduled Veterinary Services	68.4	65.8	87.5	90.1
Nutritionist to Design Feed Mixes	71.6	72.6	87.8	90.0
Individual Cow Production Records	60.6	63.6	81.6	84.9
Forward Purchasing of Inputs	19.4	21.9	44.2[c]	53.1[d]
Negotiated Price Discounts for Inputs	34.5	36.2	60.0[c]	71.2[d]
Production Systems				
Milk Cows ≥3 Times per Day	7.0	9.4	30.4[c]	41.8[d]
Dairy Parlor	49.9	53.0	83.9[c]	89.3[d]
Pasture-based Milk Production	18.3	20.0	6.8	6.6
USDA Certified Organic Milk Production	1.6	8.6	0.7[c]	3.3[d]
Controlled Breeding and/or Calving Season	25.3	24.8	24.2	20.9

Superscripts [a] and [b] indicate that the estimates differ significantly at P < 0.10. Likewise, superscripts [c] and [d] indicate that the estimates differ significantly at P < 0.10.

significant. This is compared with estimated aggregate adoption rates of 24.4% in 2000 (Khanal et al., 2010) (which differed from the 2005 estimated at $P \leq 0.10$) and 12.5% in 1993 (Short, 2000). The percentage of milk produced by farms using automatic take-offs was 72.9% and 77.0% in 2005 and 2010, respectively. It appears that significant diffusion of the technology occurred up to 2005, with little additional diffusion thereafter.

Table 3 Logit results of farmer adoption of computerized/automated technology, 2010 (Standard errors in parenthesis)

Variable	Units	Computerized Feed Delivery System		Computerized Milking System		Computer to Manage Dairy Records	
		β	Marg Effect	β	Marg Effect	β	Marg Effect
Constant		-4.8220*** (0.8249)		-5.6313*** (0.9065)		-2.7211*** (0.6024)	
Farm Structure							
Cows	No/100	0.1220*** (0.0336)	0.0074*** (0.0020)	0.0214 (0.0195)	0.0010 (0.0010)	0.4062*** (0.1406)	0.0892*** (0.0330)
Acres	No/100	0.0515** (0.0225)	0.0031** (0.0010)	0.0791*** (0.0261)	0.0038*** (0.0010)	0.1081*** (0.0334)	0.0237*** (0.0070)
Owned	Portion	-0.4698* (0.2717)	-0.0285* (0.0162)	0.2866*** (0.1052)	0.0138*** (0.0050)	0.1438 (0.1046)	0.0316 (0.0231)
Specialized	Portion	1.9100** (0.7621)	0.1161** (0.0477)	2.0391** (0.8850)	0.0983** (0.0425)	0.7340 (0.6151)	0.1612 (0.1334)
Organic	0-1	-0.9541** (0.3986)	-0.0419*** (0.0154)	-0.5936* (0.3557)	-0.0231* (0.0131)	0.0266 (0.2031)	0.0059 (0.0449)
Pasture-based	0-1	-0.0877 (0.5212)	-0.0052 (0.0301)	-0.3445 (0.5208)	-0.0152 (0.0201)	-0.2756 (0.2593)	-0.0587 (0.0527)
Farmer Demographics							
Age	Yrs/10	0.1025 (0.1009)	0.0062 (0.0062)	0.1017 (0.1078)	0.0049 (0.0051)	0.0002 (0.0684)	0.0000 (0.0150)
College	0-1	0.5915* (0.3127)	0.0440 (0.0278)	0.8527*** (0.2893)	0.0559** (0.0239)	1.0507*** (0.2400)	0.2506*** (0.0580)
Regions							
West	0-1	0.4910 (0.4646)	0.0367 (0.0408)	0.0486 (0.4873)	0.0024 (0.0244)	0.7398* (0.3949)	0.1769* (0.0975)
Pacific	0-1	0.3543 (0.4289)	0.0247 (0.0334)	0.9872** (0.3992)	0.0715* (0.0391)	0.3691 (0.4413)	-0.0587 (0.0527)
Southeast	0-1	-0.9249* (0.5238)	-0.0385* (0.0161)	-3.6464 (2.2622)	-0.0513*** (0.0088)	0.1809 (0.3427)	0.0409* (0.0789)
Corn Belt	0-1	0.3359 (0.3528)	0.0225 (0.0250)	-0.0159 (0.3860)	-0.0008 (0.0185)	0.3226 (0.2174)	0.0732 (0.0501)
Northern Plains	0-1	-0.0263 (0.5652)	-0.0016 (0.0336)	-0.6097 (0.7224)	-0.0227 (0.0209)	0.3226 (0.2174)	-0.0277 (0.0849)
Appalachia	0-1	-0.1633 (0.3950)	-0.0093 (0.0215)	-0.0257 (0.3921)	-0.0012 (0.0186)	-0.2235 (0.2494)	-0.0472 (0.0511)
Northeast	0-1	-0.1040 (0.4136)	-0.0062 (0.0241)	0.0795 (0.4258)	0.0039 (0.0212)	-0.0914 (0.2640)	-0.0199 (0.0571)
Southern Plains	0-1	-2.0439** (1.0421)	-0.0586*** (0.0137)	-3.6757 (3.0604)		-0.6904 (0.4545)	-0.1317* (0.0754)
Diagnostics							
Pseudo R-square		0.1511		0.1266		0.1989	

Table 3 Logit results of farmer adoption of computerized/automated technology, 2010 (Standard errors in parenthesis) *(Continued)*

Variable	Units	Computerized Feed Delivery System		Computerized Milking System		Computer to Manage Dairy Records	
		β	Marg Effect	β	Marg Effect	β	Marg Effect
Constant		−1.3024*** (0.5003)		−1.8703*** (0.5522)		−4.4040*** (0.9720)	
Farm Structure							
Cows	No./100	0.2420** (0.0998)	0.0605** (0.0250)	0.3545* (0.1933)	0.0877* (0.0490)	0.0369* (0.0199)	0.0012* (0.0007)
Acres	No./100	0.0820*** (0.0293)	0.0205*** (0.0070)	0.1654*** (0.0421)	0.0409*** (0.0100)	0.0090 (0.0174)	0.0002 (0.0010)
Owned	Portion	0.0285 (0.1061)	0.0071 (0.0265)	0.1133 (0.1227)	0.0280 (0.0304)	0.0779 (0.1115)	0.0025 (0.0035)
Specialized	Portion	1.0941** (0.4581)	0.2733** (0.1145)	1.1199** (0.5113)	0.2769** (0.1250)	−0.5187 (0.6899)	−0.0168 (0.0227)
Organic	0-1	0.0297 (0.1730)	0.0074 (0.0432)	−0.6630*** (0.2012)	−0.1552*** (0.0439)	−0.6928 (0.4412)	−0.0174* (0.0096)
Pasture-based	0-1	−0.3330 (0.2134)	−0.0830 (0.0528)	−0.7728*** (0.2555)	−0.1821*** (0.0547)	−0.9543* (0.5126)	−0.0243** (0.0098)
Farmer Demographics							
Age	Yrs./10	−0.0318 (0.0701)	−0.0079 (0.0175)	−0.0911 (0.0686)	−0.0225 (0.0170)	0.2504*** (0.0951)	0.0810** (0.0350)
College	0-1	1.5028*** (0.2834)	0.3286*** (0.0485)	0.8154*** (0.2822)	0.2006*** (0.0666)	0.1055 (0.2457)	0.0035 (0.0085)
Regions							
West	0-1	0.1956 (0.3883)	0.0486 (0.0957)	−1.2902** (0.6198)	−0.2677*** (0.1018)	1.2892* (0.5237)	0.0766* (0.0460)
Pacific	0-1	−0.0166 (0.3635)	−0.0042 (0.0908)	1.2490*** (0.4616)	0.2944*** (0.0972)	3.6362*** (0.4186)	0.4907*** (0.0799)
Southeast	0-1	−0.4735 (0.3227)	−0.1170 (0.0779)	−1.5584*** (0.5061)	−0.3041*** (0.0749)	2.3456*** (0.4211)	0.2279*** (0.0668)
Corn Belt	0-1	−0.3144 (0.2014)	−0.0784 (0.0500)	−0.3053 (0.2074)	−0.0744 (0.0500)	−0.2714 (0.5588)	−0.0081 (0.0158)
Northern Plains	0-1	−0.1517 (0.4114)	−0.0379 (0.1027)	0.9535 (0.4754)	0.2304** (0.1046)	0.4631 (0.8462)	0.0186 (0.0408)
Appalachia	0-1	−0.7061*** (0.2290)	−0.1719*** (0.0531)	0.5523** (0.2214)	0.1372** (0.0541)	−0.0985 (0.5255)	−0.0031 (0.0159)
Northeast	0-1	−0.6250*** (0.2307)	−0.1548*** (0.0560)	−0.2459 (0.2386)	−0.0603 (0.0579)	0.0964 (0.5536)	0.0032 (0.0186)
Southern Plains	0-1	−0.6308 (0.4073)	−0.1542 (0.0950)	−1.6702*** (0.5661)	−0.3204*** (0.0789)	1.7811*** (0.5201)	0.1327** (0.0624)
Diagnostics							
Pseudo R-square		0.1223		0.1916		0.3086	

Note: *, **, and *** indicate significance at the $P \leq 0.10$, $P \leq 0.05$, and $P \leq 0.01$ levels, respectively.

In 2010, large-scale, specialized, and educated farmers exhibited higher probability of using the technology. Furthermore, USDA certified organic or pasture-based operations had lower usage rates of 15.5 and 18.2 percentage points, respectively. Though diffusion was relatively stagnant between 2005 and 2010, increases in farm size and farmer education suggest modest diffusion in the near future.

In 2010, the aggregate adoption rate of holding pens with udder washers was 6.7% of farmers, compared with 6.5% in 2005, but the difference was not statistically significant. This is compared with estimated aggregate adoption of 5.4% in 2000 (Khanal et al., 2010) and 3.3% in 1993 (Short, 2000). The percentage of milk produced by farms using holding pens with udder washers was 31.4% and 30.7% in 2005 and 2010, respectively. In 2010, farmers with more cows were greater users and an additional 10 years of the farmer's age increased usage by 0.8 percentage points. USDA certified organic and pasture-based operations were lower users of this technology. Though diffusion of this technology has stagnated in recent years, increases in farm size and farmer education suggest modest diffusion in coming years.

Breeding and/or biological technologies

Aggregate adoption rates for breeding and/or biological technologies are shown in Figure 2 and logit results are shown in Table 4. In 2010, the aggregate adoption rate of artificial insemination was 80.1% of farmers, compared with 81.5% in 2005, but the difference was not statistically significant. Khanal et al. (2010) showed that in 2000, 64.3% of farmers used, as stated in the 2000 ARMS dairy survey, "genetic selection and breeding programs (embryo transplants, artificial insemination)." Since embryo transplants would rarely be used without artificial insemination, it is probable that all of these farms used artificial insemination, making this a reasonable estimate of the percentages of farms using the technology. Given that the 2000 and 2005 usage differed at $P \leq 0.10$, it appears that significant adoption diffusion occurred prior to 2005, but not thereafter. The percentage of milk produced by farms using artificial insemination increased from 88.9% in 2005 to 92.2% in 2010. Greater users of the technology in 2010 were large-scale, specialized, and educated farmers. On the other hand, operating a USDA certified organic or pasture-based operation reduced usage. Artificial insemination adoption may be nearing an equilibrium point since 90% of the milk produced is now covered by this technology.

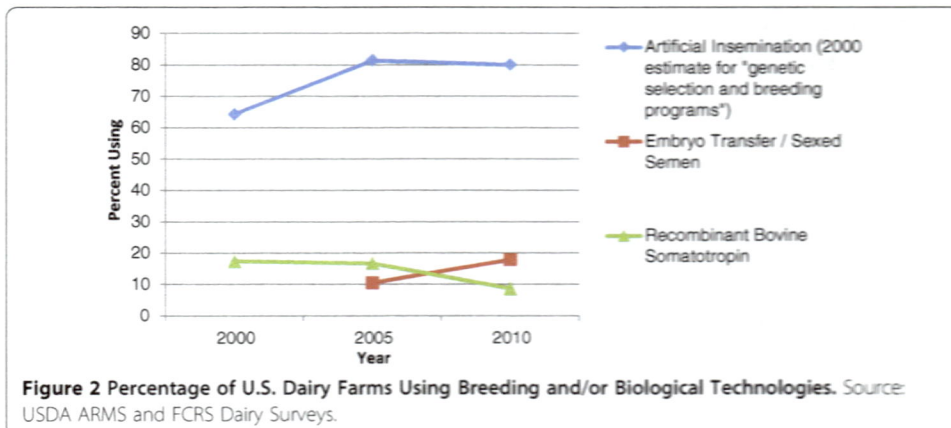

Figure 2 Percentage of U.S. Dairy Farms Using Breeding and/or Biological Technologies. Source: USDA ARMS and FCRS Dairy Surveys.

In 2010, the aggregate adoption rate of embryo transfer and/or sexed semen was 17.8% of farmers, compared with 10.4% in 2005, statistically different at $P \leq 0.05$. The percentage of milk produced by farms using embryo transfer and/or sexed semen technology increased from 15.7% to 24.3% from 2005 to 2010, respectively. In 2010, large-scale, young, and educated farmers were greater users of this technology, while USDA certified organic farmers were lower users. Strong increases are attributed primarily to the diffusion of sexed semen, with additional diffusion expected as farms become larger and more highly educated farmers enter the industry.

In 2010, the aggregate adoption rate of rbST was 8.5% of farmers compared with 16.6% in 2005, statistically different at $P \leq 0.01$. This is compared with an estimated aggregate adoption of 17.3% in 2000 (Khanal et al., 2010) and shows significant decline in the percentage of farms using rbST. This reduction is likely due to consumer concerns regarding rbST and resultant premiums for milk produced without the use of rbST. Past economic studies that have not shown increased profitability with rbST adoption also partially explain declining usage. Furthermore, to get the desired production response with rbST, greater management of feed inputs is needed. The percentage of milk produced by farms using rbST decreased from 40.0% in 2005 to 16.6% in 2010. For 2010, greater ownership of land increased rbST usage.

Management practices

Aggregate adoption rates for management practices are shown in Figure 3 and logit results are shown in Table 5. In 2010, the aggregate adoption rate for regularly scheduled veterinary services was 65.8% of farmers, compared with 68.4% in 2005, but the difference was not statistically significant. The percentages of milk produced by farms using regularly scheduled veterinary services were 87.5% and 90.1% in 2005 and 2010, respectively. In 2010, larger-scale and younger farmers were greater users, while certified organic and pasture-based farms were lower users. Increased numbers of larger-scale farms (greater use) along with increased numbers of certified organic farms (lower use) partially explain stagnant adoption.

In 2010, aggregate adoption of a nutritionist to design feed mixes was 72.6% of farmers, compared with 71.6% in 2005, but the difference was not statistically significant. This is compared with an estimated aggregate adoption rate of 66.9% in 2000 (Khanal et al., 2010). The percentage of milk produced by farmers using a nutritionist to design feed mixes was 87.8% and 90.0% in 2005 and 2010, respectively. In 2010, younger farmers and those who had more cows were greater users, while certified organic and pasture-based farmers had lower usage rates. Modest diffusion over the past decade may be partially explained by the opposing factors of increased farm size and increased certified organic production.

The 2010 aggregate adoption rate of individual cow production records was 63.6% of farmers compared with 60.6% in 2005, but the difference was not statistically significant. The percentage of milk produced by farmers keeping individual cow production records was 81.6% and 84.9% in 2005 and 2010, respectively. In 2010, large-scale, specialized, educated, and organic-certified farmers were greater users, while pasture-based farmers were lower users. Though rapid diffusion of this practice is not evident, it is generally considered by dairy farm management professionals to be an important component in increasing productivity. More larger and certified organic farms should result in greater diffusion.

Table 4 Logit results of dairy farmer adoption of breeding and/or biological technologies, 2010 (Standard errors in parenthesis)

Variable	Units	Artificial Insemination		Embryo Transfer and/or Sexed Semen		Recombinant Bovine Somatotropin	
		β	Marg Effect	β	Marg Effect	β	Marg Effect
Constant		0.5505 (0.5881)		−1.1789** (0.5862)		−2.1980*** (0.9552)	
Farm Structure							
Cows	No./100	0.0904** (0.0039)	0.0124 (0.0050)	−0.0080 (0.0137)	−0.0011 (0.0020)	0.0227 (0.0156)	0.0014 (0.0010)
Acres	No./100	−0.0026 (0.0203)	−0.0004 (0.0030)	0.0430*** (0.0150)	0.0058*** (0.0020)	0.0759*** (0.0258)	0.0047*** (0.0020)
Owned	Portion	0.0272 (0.0947)	0.0037 (0.0130)	0.1152 (0.1086)	0.0156 (0.0146)	0.2500** (0.1140)	0.0153*** (0.0069)
Specialized	Portion	1.3565*** (0.5033)	0.1857*** (0.0704)	0.2655 (0.4738)	0.0359 (0.0642)	0.8526 (0.7443)	0.0523 (0.0458)
Organic	0-1	−1.0418*** (0.1950)	−0.1848*** (0.0378)	−1.7048*** (0.2730)	−0.1428*** (0.0189)	−0.5205 (0.4887)	−0.0280 (0.0224)
Pasture-based	0-1	−0.6382*** (0.2245)	−0.0990*** (0.0385)	−0.2700 (0.3065)	−0.0345 (0.0367)		
Farmer Demographics							
Age	Yrs./10	0.0585 (0.0842)	0.0080 (0.0116)	−0.1524* (0.0834)	−0.0206* (0.0112)	−0.2132 (0.1448)	−0.0131 (0.0089)
College	0-1	0.9075*** (0.2945)	0.0979*** (0.0243)	0.7171*** (0.2473)	0.1156** (0.0456)	0.3967 (0.3398)	0.0278 (0.0265)
Regions							
West	0-1	−1.3309*** (0.4031)	−0.2591*** (0.0948)	−0.5137 (0.3592)	−0.0584* (0.0351)	−3.1607*** (1.2306)	−0.0663*** (0.0104)
Pacific	0-1	−0.2934 (0.3836)	−0.0438 (0.0618)	0.1479 (0.3170)	0.0209 (0.0463)	−1.2464** (0.5305)	−0.0488*** (0.0142)
Southeast	0-1	−2.0875*** (0.3188)	−0.4461*** (0.0723)	−0.5125 (0.4233)	−0.0581 (0.0408)		
Corn Belt	0-1	−0.9362*** (0.2398)	−0.1550*** (0.0445)	−0.0845 (0.2577)	−0.0112 (0.0338)	−1.0171*** (0.3797)	−0.0481*** (0.0152)
Northern Plains	0-1	−0.6402 (0.4731)	−0.1066 (0.0917)	−0.2606 (0.4214)	−0.0323 (0.0481)	−0.7839 (0.5455)	−0.0349* (0.0179)
Appalachia	0-1	−1.9202*** (0.2434)	−0.3981*** (0.0543)	−0.1872 (0.2756)	−0.0239 (0.0337)	−2.2961*** (0.7466)	−0.0647*** (0.0115)
Northeast	0-1	−0.0168 (0.2927)	−0.0023 (0.0403)	0.0724 (0.2900)	0.0099 (0.0400)	−0.2535 (0.3408)	−0.0148 (0.0189)
Southern Plains	0-1	−2.5655*** (0.3651)	−0.5501*** (0.0733)	−1.7184** (0.6979)	−0.1319*** (0.0271)	−2.6556* (1.5891)	−0.0643*** (0.0126)
Diagnostics							
Pseudo R-square		0.1217		0.0459		0.0892	

Note: *, **, and *** indicate significance at the P ≤0.10, P ≤0.05, and P ≤0.01 levels, respectively.

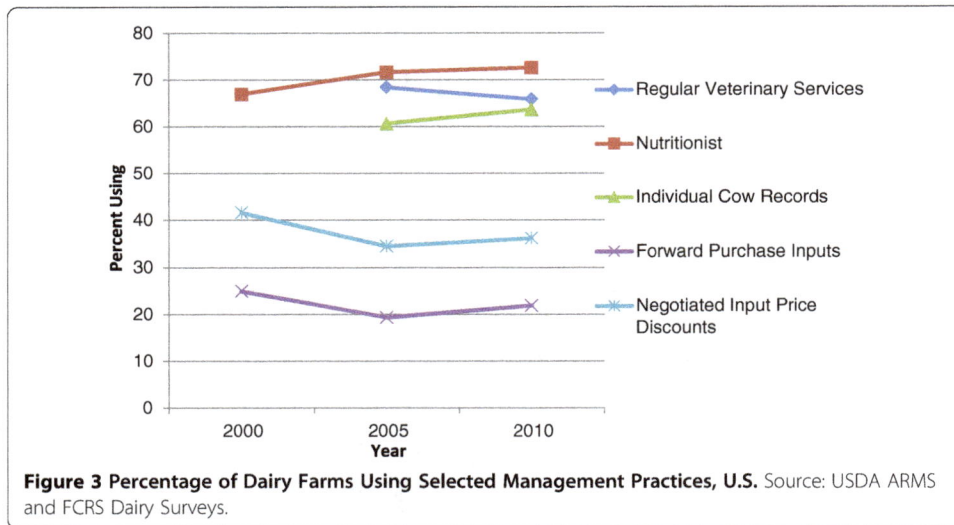

Figure 3 Percentage of Dairy Farms Using Selected Management Practices, U.S. Source: USDA ARMS and FCRS Dairy Surveys.

In 2010, 21.9% of farmers forward-purchased inputs, compared with 19.4% in 2005, but the difference was not statistically significant. This is compared with 25.0% of farmers forward purchasing inputs in 2000. The percentage of milk produced by farmers forward-purchasing inputs increased from 44.2% in 2005 to 53.1% in 2010. In 2010, farmers who were larger-scale, younger, educated, and those who owned greater percentages of their farmland were greater users, while certified organic and pasture-based producers were lower users. Though diffusion of this practice is not noticeable on a percentage-of-farms basis, modest diffusion is expected with the entry of larger-scale farms with more highly educated producers.

In 2010, 36.2% of farmers negotiated price discounts for inputs, compared with 34.5% in 2005, but the difference was not statistically significant. This is compared with 41.6% of farmers negotiating price discounts in 2000. The percentage of milk produced by farmers who negotiated for price discounts for inputs increased from 60.0% in 2005 to 71.2% in 2010. In 2010, larger-scale, younger, and educated farmers, and those who owned greater percentages of their farmland were greater users, while certified organic farmers were lower users. Implications for further diffusion of this management practice are similar to those for forward purchasing of inputs.

Production systems

Aggregate adoption rates for production systems are shown in Figure 4 and logit results are shown in Table 6. In 2010, 9.4% of farmers milked cows at least 3 times a day, compared with 7.0% in 2005, but the difference was not statistically significant. This is compared with 3.4% of farmers milking at least 3 times per day in 2000 (Khanal et al., 2010) (which differed from the 2005 estimate at $P \leq 0.10$) and 2.7% in 1993. Overall, the percentage of farmers milking cows at least 3 times daily has increased modestly over the past two decades. The percentage of milk produced by farms milking cows at least 3 times per day increased from 30.4% in 2005 to 41.8% in 2010. In 2010, larger-scale, specialized, and more educated farmers were greater users, while certified organic and pasture-based farmers were lower users. Though the differences in the percentage of farms using this system was not statistically significant over the period, 2005–2010, it

Table 5 Logit results for dairy farmer adoption of management practices, 2010 (Standard errors in parenthesis)

Variable	Units	Regular Veterinary Services		Nutritionist to Design Feed Mixes/Purchase		Keep Individual Cow Production Records		Forward Purchasing of Inputs		Negotiate Price Discounts for Inputs	
		β	Marg Effect	β	Marg Effect	β	Marg Effect	β	Marg Effect	β	Marg Effect
Constant		0.9779* (0.5619)		2.2682*** (0.5681)		-0.0730 (0.5371)		-0.9932 (0.6556)		-0.2132 (0.5356)	
Farm Structure											
Cows	No./100	0.3171** (0.1491)	0.0638** (0.0270)	0.1747* (0.0907)	0.0319** (0.0160)	0.1170 (0.0874)	0.0263 (0.0190)	0.0969** (0.0392)	0.0152** (0.0060)	0.2356*** (0.0616)	0.0555*** (0.0150)
Acres	No./100	0.0807*** (0.0290)	0.0162*** (0.0060)	0.0245 (0.0230)	0.0045 (0.0040)	0.0492** (0.0241)	0.0111** (0.0060)	0.0670*** (0.0218)	0.0110*** (0.0030)	0.0769*** (0.0235)	0.0181*** (0.0050)
Owned	Portion	0.1526 (0.0932)	0.0307 (0.0187)	0.0129 (0.1027)	0.0024 (0.0188)	0.1515* (0.0906)	0.0341* (0.0204)	0.1977* (0.1057)	0.0310* (0.0167)	0.1737* (0.0896)	0.0409* (0.0212)
Specialized	Portion	0.3708 (0.4797)	0.0745 (0.0971)	0.2184 (0.5011)	0.0399 (0.0915)	1.0345** (0.4783)	0.2329** (0.1084)	0.1089 (0.5634)	0.0171 (0.0883)	0.1536 (0.4778)	0.0362 (0.1125)
Organic	0-1	-0.7560*** (0.1825)	-0.1696*** (0.0437)	-0.8361*** (0.1802)	-0.1771*** (0.0403)	0.3477* (0.1894)	0.0744* (0.0390)	-0.6341*** (0.2225)	-0.0841*** (0.0265)	-0.6024*** (0.1830)	-0.1312*** (0.0376)
Pasture-based	0-1	-1.4377*** (0.2251)	-0.3250*** (0.0546)	-1.2251*** (0.2219)	-0.2590*** (0.0518)	-1.0914*** (0.2155)	-0.2599*** (0.0521)	-1.2014*** (0.2702)	-0.1504*** (0.0262)	-0.2873 (0.2310)	-0.0661 (0.0516)
Farmer Demographics											
Age	Yrs/10	-0.1978*** (0.0767)	-0.0398*** (0.0154)	-0.2680*** (0.0732)	-0.0489*** (0.0132)	-0.0994 (0.0690)	-0.0224 (0.0155)	-0.1863** (0.0793)	-0.0292** (0.0126)	-0.2208*** (0.0697)	-0.0520*** (0.0165)
College	0-1	0.1120 (0.2441)	0.0021 (0.0474)	0.1344 (0.2611)	0.0239 (0.0452)	0.6260** (0.3020)	0.1285** (0.0553)	0.9036*** (0.2394)	0.1709*** (0.0518)	0.5234** (0.2171)	0.1275** (0.0573)
Regions											
West	0-1	0.1626 (0.3951)	0.0315 (0.0746)	-0.1990 (0.3567)	-0.0381 (0.0707)	-0.4782 (0.3339)	-0.1139 (0.0824)	0.4811 (0.3260)	0.0861 (0.0781)	-0.0821 (0.3375)	-0.0192 (0.0781)
Pacific	0-1	0.1620 (0.3884)	0.0315 (0.0738)	-0.5675* (0.3451)	-0.1163 (0.0760)	0.3482 (0.3762)	0.0742 (0.0763)	0.2478 (0.3260)	0.0415 (0.0576)	-0.0817 (0.3208)	-0.0191 (0.0743)

Table 5 Logit results for dairy farmer adoption of management practices, 2010 (Standard errors in parenthesis) *(Continued)*

Southeast	0-1	-1.3025***	-0.3077***	-0.5115	-0.1048	0.2343	0.0507	-0.9940**	-0.1135***	-0.3641	-0.0815
		(0.3561)	(0.0841)	(0.3379)	(0.0746)	(0.3454)	(0.0721)	(0.3960)	(0.0327)	(0.3387)	(0.0718)
Corn Belt	0-1	-0.1350	-0.0277	-0.2997	-0.0574	-0.1171	-0.0267	0.0895	0.0143	-0.2650	-0.0610
		(0.2153)	(0.0447)	(0.2263)	(0.0447)	(0.2005)	(0.0460)	(0.2222)	(0.0359)	(0.2030)	(0.0459)
North Plains	0-1	-1.0329**	-0.2411**	-0.3279	-0.0647	0.0725	0.0161	0.7164*	0.1358	-0.4551	-0.1003
		(0.4103)	(0.1026)	(0.3884)	(0.0815)	(0.3611)	(0.0795)	(0.3945)	(0.0860)	(0.3930)	(0.0801)
Appalachia	0-1	-0.0493	-0.0100	-0.6199**	-0.1285**	-0.0926	-0.0211	0.1971	0.0326	0.1571	0.0376
		(0.2451)	(0.0501)	(0.2457)	(0.0546)	(0.2250)	(0.0518)	(0.2412)	(0.0415)	(0.2177)	(0.0526)
Northeast	0-1	0.2990	0.0583	0.3519	0.0616	-0.0007	-0.0002	-0.1123	-0.0173	-0.1129	-0.0264
		(0.2595)	(0.0488)	(0.2634)	(0.0441)	(0.2407)	(0.0542)	(0.2859)	(0.0433)	(0.2379)	(0.0553)
South Plains	0-1	-1.8523***	-0.4322***	-1.1208***	-0.2503***	-0.8597**	-0.2087**	-0.2819	-0.0406	-0.4848	-0.1064
		(0.4339)	(0.0898)	(0.3858)	(0.0939)	(0.3474)	(0.0848)	(0.4069)	(0.0539)	(0.3829)	(0.0775)
Diagnostics											
Pseudo R^2		0.1510		0.1141		0.0750		0.1214		0.1038	

Note: *, **, and *** indicate significance at the $P \leq 0.10$, $P \leq 0.05$, and $P \leq 0.01$ levels, respectively.

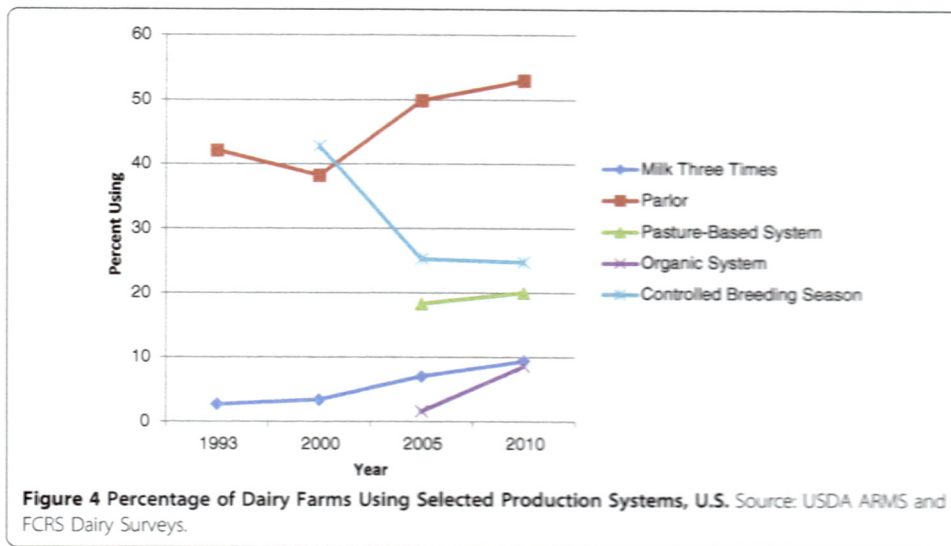

Figure 4 Percentage of Dairy Farms Using Selected Production Systems, U.S. Source: USDA ARMS and FCRS Dairy Surveys.

appears that over the past two decades, steady diffusion has continued with larger-scale, more highly educated producers entering the industry.

In 2010, the aggregate adoption rate of a milking parlor was 53.0% of farmers, compared with 49.9% in 2005, but the difference was not statistically significant. This is compared with an aggregate adoption rate of 38.2% in 2000 (Khanal et al., 2010) (which differed from the 2005 percentage at $P \leq 0.10$) and 42.1% in 1993 (Short 2000). Overall, the percentage of farms using parlors increased over the past two decades. The percentage of milk produced by farms under a parlor system increased from 83.9% to 89.3% from 2005 to 2010, respectively. In 2010, large-scale farmers and those who operated pasture-based operations were greater users. Though diffusion has been modest but steady over the past two decades, the striking increase in adoption with more cows suggests that new, larger entrants will be adopters.

Logit models were not estimated for pasture-based milk production or certified organic milk production because, while they are production systems, they are used as independent variables in the other TMPPS models and are, thus, considered TMPPS adoption drivers. We do, however, examine usage rates for these production systems. In 2010, 20.0% of operations were pasture-based, compared with 18.3% in 2005, but the difference was not statistically significant. Percentages of milk produced by farms under pasture-based systems were 6.8% and 6.6% in 2005 and 2010, respectively. Farmers in some markets may receive premiums for pasture-based milk, one of the reasons being that milk produced from pasture-based dairies may have lower somatic cell counts along with higher protein and butterfat components (Horner et al. 2012). Furthermore, USDA certified organic production, which requires cows to have access to pasture, has increased. These factors are likely the major drivers allowing aggregate adoption of pasture-based production to hold steady. Though pasture-based production has not shown significant diffusion in recent years, future consumer demand for milk produced under more "sustainable" systems could be a significant driver influencing its use in coming years.

In 2010, 8.6% of the operations were certified organic, compared with 1.6% in 2005, but the difference was not statistically significant, as the coefficients of variation for the estimates were relatively high for both years. The percentage of milk produced by farms using

Table 6 Logit results for dairy farmer adoption of production systems, 2010 (Standard errors in parenthesis)

Variable	Units	Milk Cows ≥3 Times/Day		Use of a Dairy Parlor		Control Breeding and/or Calving Season	
		β	Marg Effect	β	Marg Effect	β	Marg Effect
Constant		−6.8569*** (1.3940)		−2.9046*** (0.7228)		1.2252* (0.5979)	
Farm Structure							
Cows	No./100	0.1168*** (0.0427)	0.0053** (0.0020)	3.2069*** (0.4211)	0.6800*** (0.0610)	−0.0097 (0.0135)	−0.0017 (0.0020)
Acres	No./100	0.0853*** (0.0261)	0.0039*** (0.0010)	0.0358 (0.0516)	0.0076 (0.0110)	−0.0229 (0.0166)	−0.0041 (0.0030)
Owned	Portion	0.3359*** (0.1154)	0.0152*** (0.0051)	−0.1253 (0.0974)	−0.0266 (0.0207)	−0.3062* (0.1670)	−0.0545* (0.0296)
Specialized	Portion	4.5926*** (1.3418)	0.2084*** (0.0617)	−0.7025 (0.6383)	−0.1490 (0.1336)	−0.2972 (0.4909)	−0.0529 (0.0873)
Organic	0-1	−2.6073*** (0.9107)	−0.0542*** (0.0114)	0.1841 (0.2108)	0.0378 (0.0421)	0.5651*** (0.1937)	0.1126*** (0.0400)
Pasture-based	0-1	−1.2610 (0.7799)	−0.0425*** (0.0170)	0.5428** (0.2322)	0.1072*** (0.0418)	0.7396*** (0.2317)	0.1463*** (0.0502)
Farmer Demographics							
Age	Yrs/10	0.0159 (0.1527)	0.0007 (0.0069)	0.0482 (0.0987)	0.0102 (0.0210)	0.0424 (0.0694)	0.0075 (0.0123)
College	0-1	1.1874*** (0.2842)	0.0838*** (0.0277)	0.0248 (0.3723)	0.0052 (0.0783)	0.2968 (0.2374)	0.0560 (0.0472)
Regions							
West	0-1	0.0407 (0.4535)	0.0019 (0.0213)			0.5220* (0.3079)	0.1048 (0.0668)
Pacific	0-1	−1.1548** (0.5672)	−0.0341*** (0.0119)			0.3633 (0.2941)	0.0701 (0.0601)
Southeast	0-1	−2.3947*** (0.5348)	−0.0442*** (0.0087)			1.6935*** (0.3116)	0.3884*** (0.0713)
Corn Belt	0-1	−0.8453*** (0.3685)	−0.0306*** (0.0126)	2.0361*** (0.2759)	0.3186*** (0.0367)	0.4468** (0.2277)	0.0855* (0.0450)
North Plains	0-1	−0.0042 (0.5353)	−0.0002 (0.0242)			1.3877*** (0.3852)	0.3141*** (0.0936)
Appalachia	0-1	−0.9534** (0.3915)	−0.0299*** (0.0101)	3.2609*** (0.3809)	0.3231*** (0.0361)	0.9836*** (0.2441)	0.2114*** (0.0561)
Northeast	0-1	−0.5939 (0.3898)	−0.0242 (0.0148)	−0.1933 (0.2540)	−0.0415 (0.0555)	−0.1774 (0.2772)	−0.0309 (0.0472)
South Plains	0-1	−1.5348** (0.7694)	−0.0382*** (0.0115)			0.9363*** (0.3637)	0.2015*** (0.0870)
Diagnostics							
Pseudo R²		0.2374		0.3665		0.0569	

Note: *, **, and *** indicate significance at the P ≤0.10, P ≤0.05, and P ≤0.01 levels, respectively.

a certified organic system increased from 0.7% in 2005 to 3.3% in 2010. Further diffusion of certified organic milk production will likely be driven by consumer demand.

In 2010, 24.8% of the farmers controlled their breeding and/or calving seasons, compared with 25.3% in 2005, but the difference was not statistically significant. This is compared with 42.8% of farmers controlling breeding seasons in 2000. The percentages of milk produced by farms controlling breeding and/or calving seasons were 24.2% and 20.9% in 2005 and 2010, respectively. In 2010, small-scale and certified organic farmers, and those who operated pasture-based systems were greater users. Though use of this system appears to be declining, future usage may depend partially on the usage of pasture-based and certified organic systems.

Adoption relationships among TMPPS

Previous research has established that adopters of most TMPPS are also adopters of most of the others, with Khanal et al. (2010) and Pruitt et al. (2012) showing this result for the U.S. dairy and cow-calf industries, respectively. Our research supports these findings, showing that most TMPPS are technically complementary or, at the very least, adopters of one TMPPS are prone to also adopt most others. We used the Fisher exact test (Zar, 1984, p. 390) to determine whether usage among TMPPS was correlated and, in cases where correlation was found, the Cramer coefficient (Zar, 1984, p. 322) to determine whether a complementary or substitute relationship existed among each of the TMPPS. These analyses were conducted using the 2010 ARMS dairy data with the provided weights. Results showed that adoption of all TMPPS was positively correlated with the adoption of all other TMPPS with the exception of: (1) use of a certified organic system was negatively correlated with all other TMPPS except for use of a breeding season and a pasture-based system; (2) use of a pasture-based system was negatively correlated with all other TMPPS except for uses of a breeding season and a certified organic system; (3) use of a breeding season was negatively correlated with veterinary services, use of a nutritionist, forward purchasing of inputs, three-times daily milking, a holding pen with an udder washer, and rbST; (4) use of artificial insemination was negatively correlated with uses of a certified organic system, a pasture-based system, and a parlor; and (5) no relationships were found between artificial insemination and breeding season, parlor and embryo transfer / sexed semen, and rbST and holding pen with an udder washer. Overall, results show striking support for previous findings suggesting that TMPPS adopters tend to be adopters of other TMPPS except for cases of alternative systems such as certified organic and pasture-based production.

Conclusions

Aggregate adoption rates of most major productivity-enhancing TMPPS have increased over the past couple of decades. Of the computerized and/or automated technologies, automatic take-offs have experienced the highest adoption diffusion rates, moving from 13% to 40% aggregate adoption from 1993 to 2010, though diffusion appears to have slowed during 2005–2010. Most of the other technologies in this category have experienced relatively slow adoption diffusion, with all but a computer for record-keeping and internet use for dairy information having aggregate adoption rates less than 10% throughout the period of study. It is noted, however, that for three of these technologies, percentages of milk produced covered by the technologies greatly exceeds percentages of farms using them, suggesting greater usage among larger farms. This is further supported by logit results showing farm size to be positively associated with usage of the six technologies in this category. Furthermore, farms more heavily specialized in milk production and farmers with college degrees were the more likely users of these technologies while certified organic and pasture-based farms were generally the less likely users.

Of the three breeding and/or biological technologies, only artificial insemination held steady during the 2005–2010 period: 80-82% of farms used it and 89-91% of gross value of milk production was covered by the technology. Its aggregate adoption increased relative to 2000. This technology is likely approaching a "ceiling" or equilibrium value, a result that should not be surprising given the rapid early adoption of this technology in the 1940s and its relatively high current usage. Newer technologies, such as embryo transfer / sexed semen and rbST have experienced significant diffusion, with the former increasing both in usage and percentage of production coverage by about 70% during 2005–2010. This is likely attributed primarily to sexed semen, a relatively new technology that has been expected to experience significant diffusion (DeVries et al., 2008). On the other hand, after initially modest diffusion following its commercial release in 1994, rbST usage decreased from 2005 to 2010, with a 49% reduction in farms using it. This rather dramatic decrease is likely explained by negative consumer reactions to milk produced using the technology, premiums in some cases being paid for milk not produced using rbST, and small or nonexistent impacts of rbST use on farm profitability (Tauer 2009; Gillespie et al., 2010). For each of the breeding and/or biological technologies, usage was concentrated among larger-scale operations, as indicated by the higher percentages of milk production covered by the technologies relative to percentages of farms using them, and significant *Cows* or *Acres* coefficients in the logit models. Certified organic farms were less likely to use any of these technologies (they are barred from using rbST), and more highly educated producers were more extensive users of the breeding technologies.

Each of the dairy management practices had relatively steady aggregate adoption rates during 2005–2010; percentages of farms using the practices did not differ significantly for the two years. Figure 3 illustrates a relatively steady aggregate adoption since 2005. Larger farms were greater users, as indicated by the higher percentages of milk production covered by the management practices relative to percentages of farms adopting, and significant *Cows* or *Acres* coefficients in the logit models. Increased percentages of owned land and the farmer holding a college degree increased usage, while older farmers were generally lower users of management practices. Certified organic and pasture-based producers were lower users of the management practices, with the exception of certified organic producers keeping individual cow production records, as expected due to the stringent requirements of USDA certified organic production.

Several trends can be seen in the selection of dairy systems. As farms have increased in size and dairy farming has become more intensive (versus extensive) in nature, higher percentages of farms have adopted parlor systems and are milking cows at least three times a day. In the face of these trends, however, percentages of farms producing under pasture-based systems have remained steady and certified organic dairy production has increased. This suggests that U.S. dairy production is becoming more diverse in its use of production systems, more intensive on the one hand and more likely certified organic and/or pasture-based on the other. Larger, more specialized farms operated by farmers with college degrees were more likely to milk cows at least three times a day, and larger farms were more likely to utilize a dairy parlor. Certified organic and pasture-based operations were more likely to control breeding and/or calving seasons.

It is worthwhile to point out the major impact that education had on TMPPS use. *College* was significant with a positive sign for 11 of the 17 TMPPS for which logit models were estimated. Numbers of percentage points by which usage by college degree-holding farmers exceeded that by non-degree holders ranged from six (computerized milking system) to 33 (accessing the internet for dairy information). College was particularly important for adopting computerized and/or automated technologies and breeding and/or biological technologies, underscoring the importance of training in adopting productivity-enhancing technology.

In order for many extension economists to serve their full dairy clientele, they need to be knowledgeable about the available TMPPS in terms of costs and benefits associated with each one of them. Further research should continue to focus on the advantages and disadvantages of each of these TMPPS systems under different production environments, as well as their associated costs and benefits.

Competing interests
The authors declare that they have no competing interests.

Authors' contributions
JG conducted the logic analyses and figures and drafted the manuscript. RN conducted the difference in means analysis and assisted with drafting the manuscript. IS assisted with drafting the manuscript. All authors read and approved the final manuscript.

Acknowledgement
The views expressed are those of the authors and do not necessarily represent the views or policies of the USDA.

Author details
[1]Department of Agricultural Economics and Agribusiness, Louisiana State University Agricultural Center, Martin D Woodin Hall, Baton Rouge, LA 70803, USA. [2]Economic Research Service, 1800 M St. NW, Washington, DC 20036, USA. [3]Department of Agricultural Economics and Agribusiness, Louisiana State University, Martin D Woodin Hall, Baton Rouge, LA 70803, USA.

References
Amos HE, Kiser T, Lowenstein M (1985) Influence of milking frequency on productive and reproductive efficiencies of dairy cows. J Dairy Sci 68:732–739
Barber KA (1983) Maximizing the impact of dairy and beef bulls through breeding technology. J Dairy Sci 66:2661–2671
Cochrane WW (1958) Farm Prices: Myth and Reality North Central Publishing, St. Paul, MN
DePeters EJ, Smith NE, Acedo-Rico J (1985) Three or two times daily milking of older cows and first lactation cows for the entire lactation. J Dairy Sci 68:123–132
DeVries A, Overton M, Fetrow J, Leslie K, Eicker S, Rogers G (2008) Exploring the impact of sexed semen on the structure of the dairy industry. J Dairy Sci 91:847–856
Dubman, R.W. (2000) Variance estimation with USDA's farm costs and returns surveys and agricultural resource management study surveys. Staff paper AGES 00–01, USDA-ERS; Washington, DC
Erdman RA, Varner M (1995) Fixed yield responses to increased milking frequency. J Dairy Sci 78:1199–1203
Feder G, Just RE, Zilberman D (1985) Adoption of agricultural innovations in developing countries. Econ Dev Cult Change 33:255–298
Foote RH (1996) Review: Dairy cattle reproductive physiology research and management - Past progress and future prospects. J Dairy Sci 79:980–990
Gillespie J, Mark T, Sandretto C, Nehring R (2009a) Computerized technology adoption among farms in the U.S. dairy industry. J Am Soc Farm Managers Rural Apprais 31:201–209
Gillespie J, Nehring R, Hallahan C, Sandretto C (2009b) Pasture-based dairy systems: Who are the producers and are their operations more profitable than conventional dairies? J Agric Resour Econ 34:412–427
Gillespie J, Nehring R, Hallahan C, Sandretto C, Tauer L (2010) Adoption of bovine somatotropin in the United States and its influence on farm profitability, 2000 and 2005. AgBioForum 13(3):251–262
Gisi DD, DePeters EJ, Pelissier CL (1986) Three times daily milking of cows in California dairy herds. J Dairy Sci 69:863
Greene WH (2000) *Econometric Analysis*. Prentice Hall, New Jersey
Griliches Z (1957) Hybrid corn: an exploration in the economics of technological change. Econometrica 25(4):501–522
Grisham E, Gillespie J (2008) Record-keeping technology adoption among dairy farmers. J Am Soc Farm Managers Rural Apprais 30:16–27

11

Hawai'i's food consumption and supply sources: benchmark estimates and measurement issues

Matthew K Loke[1,2*] and PingSun Leung[1]

* Correspondence: loke@hawaii.edu
[1]Department of Natural Resources and Environmental Management, University of Hawai'i at Mānoa, 1910 East-West Road, Sherman 101, Honolulu, Hawai'i 96822, USA
[2]Hawai'i Department of Agriculture, Agricultural Development Division, 1428 South King Street, Honolulu, Hawai'i 96814, USA

Abstract

At the current time, Hawai'i lacks an established set of benchmark estimates on the availability of food for market consumption and its supply sources. This paper serves to fill a persistent gap in the existing literature by providing an estimation framework to map the existing food supply flows from various sources and to determine the various levels of food consumption in Hawai'i. The authors suggest modified measures of food self-sufficiency and import dependency to provide a more accurate assessment on the extent of food localization in Hawai'i. The analytical framework presented in this paper can be applied to other small, open (island or regional) economies with a food localization agenda, as it provides a more discrete and appropriate set of measurements, as well as offering the lessons gained through Hawai'i's experience and challenges in the data-collating process.

Local production and imports (continental United States and foreign countries) of consumable food in Hawai'i are estimated at just over 1.14 million tonnes in 2010. Food exports totaled 175.5 thousand tonnes, leaving total available food for consumption locally at 966.6 thousand tonnes. On a *de facto* basis, per capita food consumption in Hawai'i is estimated at 657.9 kilograms in 2010. At the food group level, fresh vegetables lead with per capita food consumption of 84.2 kilograms, followed by other proteins at 69.1 kilograms, fresh fruits at 67.7 kilograms, fresh milk at 62.9 kilograms, and rice at 27.9 kilograms.

The analysis indicates that Hawai'i has an overall food self-sufficiency ratio (SSR) of 15.7% and an overall food import dependency ratio (IDR) of 102.5%. While it appears counterintuitive that the IDR exceeds 100%, this figure actually indicates the existence of food imports into Hawai'i that are then turned around and re-exported to other markets. With application of the more accurate localization ratio (LR), we estimate that only 11.6% of available food for consumption in Hawai'i was actually sourced from local production in 2010. Likewise, the modified import dependency ratio (MIDR) indicates that an estimated 88.4% of available food in Hawai'i was sourced from imports.

Keywords: Food consumption; Supply sources; Benchmark estimates; Food self-sufficiency; Food localization; Import dependency

Background

In the aftermath of the *Great Recession* of 2007–2009, public concern, interest, and debate on food security and food self-sufficiency has intensified in Hawai'i and elsewhere in the United States. This concern is real and understandable, considering Hawai'i's geographic isolation in the Central Pacific Ocean, looming threats of global warming and climate change, and the 2008 food crisis, which showed serious vulnerabilities in the global food system. Hawai'i's supply of food, as it presently exists, is vulnerable to disruptions in the shipping chain, production fluctuations in the continental United States, severe weather conditions, and sudden spikes in the prices of food products, as well as higher prices for fuel, feed, fertilizers, and other agricultural "inputs." During the 2008 food crisis, the surge in food price inflation worldwide was caused primarily by rising oil prices, depreciating U.S. dollar, increasing demand for biofuels, and export restrictions imposed by leading food producing countries (Heady and Fan 2008). In some quarters, there is a genuine desire to dissociate local food prices from the rising global oil price.

In order to start a meaningful discussion on food security or food self-sufficiency in Hawai'i, we must have a reasonably good assessment of the consumption level of food groups, the characteristics of food consumed, and its origins or supply sources. In 1937, H.H. Warner, then Director of the Agricultural Extension Service, Territory of Hawai'i, wrote on the character and variety of foods consumed by people on the Islands. He described the unique situation in Hawai'i: "Probably nowhere else in the world is there to be found a group of similar racial proportions with as distinctly varied diet habits, isolated from a large part of their natural food supply."[a] While changing diets have evolved since then, including the creation of Euro-Asian cuisine and the Hawai'i Regional Cuisine movement, Warner's insightful comment is still valid and relevant today.

At the current time, Hawai'i lacks an established set of benchmark estimates on the availability of food for market consumption and its supply sources. This paper serves to fill that persistent gap in the existing literature and proposes to present available facts and a logical empirical methodology to establish definitive estimates on various food groups consumed from local and import sources. Hence, the two primary objectives in this paper are as follows: (1) map existing food supply flows and to determine the various levels of food consumption in Hawai'i; and (2) suggest modified measures of food self-sufficiency and import dependency to provide a more accurate assessment on the extent of food localization in Hawai'i.

The amount of food consumed in Hawai'i that is sourced from imports is not readily available. This deficiency is due to the difficulties in reconciling the various data sources on food imports and food expenditures. For example, interstate trade flow data are rather rudimentary, unlike customs data on foreign imports, which are fairly disaggregated. The problem is further complicated by the conversion of data consistently from the various sources to a common baseline in the supply chain (Leung and Loke 2008).

Despite these challenges, various local studies have been conducted. A Rocky Mountain Institute study estimated the import share of food in the County of Hawai'i at 85% (Page et al. 2007). Likewise, popular food system analyst, Ken Meter, estimated that more than 90% of Hawai'i's food is imported (Halweil 2004). Later, the Ulupono Initiative estimated that Hawai'i consumers spend only 8% of their food budget on locally grown food, while spending the rest on imports (Ulupono Initiative 2011). Then, the University of Hawai'i at Hilo produced a Hawai'i Island food self-sufficiency scorecard that estimated the percentage of locally produced food consumed by commodity group, ranging from 0% for grains to 95% for fresh milk (Melrose and Delparte 2012). Recently, the state planning office, in a report on "increased food security and food self-sufficiency strategy," noted that 85–90% of Hawai'i's food is imported (OP-DBEDT 2012). These estimates have been widely cited despite lingering questions about and critiques of their methodologies and estimated parameters. In comparison, the food self-sufficiency measure for the New England states is estimated at 27% in 1997 (Holm et al. 2000).

Methods

Data requirements and food groups

The apparent consumption or total supply of food availability in Hawai'i is defined as local food production plus food imports (continental United States and foreign countries) less food exports (continental United States and foreign countries). Likewise, per capita food supply is assumed to be identical to per capita food consumption in the local market. All metrics and statistics in this paper refer to food available for human consumption in product weight (kilograms). While there are alternate unit measures, such as the dollar value, calorie value, and nutritional value, we nevertheless invoked *Occam's razor* by utilizing the weight measure to keep the overall analysis as simple as possible, without distorting reality or sacrificing accuracy. The measure of food self-sufficiency in dollar value could contribute to unintended results that are counterintuitive to the overall concept itself. Beyond this, one pound of prime beef steak is clearly worth more in dollar value, calorie value, and nutritional value than one pound of rice. Higher-quality products also cost more in monetary value for the same weight measure. For example, a pound of beef steak with a *USDA Prime* label will cost more than a pound of beef steak labeled *USDA Choice* or *USDA Select*. And finally, most waterborne shipping data are published in weight measures, making weight the obvious choice in the data collating process.

We emphasize that it is not the use of calorie value and nutrition value is inappropriate or unsophisticated; it is that the adoption of calorie and nutrition measures may differentiate food commodities unnecessarily, and adds levels of complexity to the entire estimating process. For example, equivalent weight measures of fresh asparagus, frozen asparagus, cooked asparagus, and canned asparagus will have varying levels of calorie and nutrition. It is possible to make the same argument for the different varieties of asparagus (white, green, purple, and wild). Furthermore, other researchers have argued that where and how the produce is grown (in fertile versus arid lands; using various cultivation practices), processed, transported, stored, and prepared will influence calorie and nutrition values.

This paper adopts the five major food groups – dairy, grains, protein, fruits, and vegetables, as defined in the USDA My Plate concept (see Figure 1). However, the aggregate estimate for overall food consumption in Hawai'i includes one additional residual food group, which includes oils and fats, sweeteners, and others less discernible food subgroups. No beverage products (e.g., soft drinks, liquor, coffee, tea and water) are included. The consumption estimates of food groups for the entire state of Hawai'i (not by county or island) are presented in total consumption and per capita consumption in weight measure. To estimate per capita consumption, the total weight estimate is divided by the *de facto* population in Hawai'i as of 2010, which takes into account, residents, stationed military personnel and dependents, and tourists visiting in the state.

Given the amount of available food in total and the components sourced from local production and imports, we can estimate the extent of total available food that is

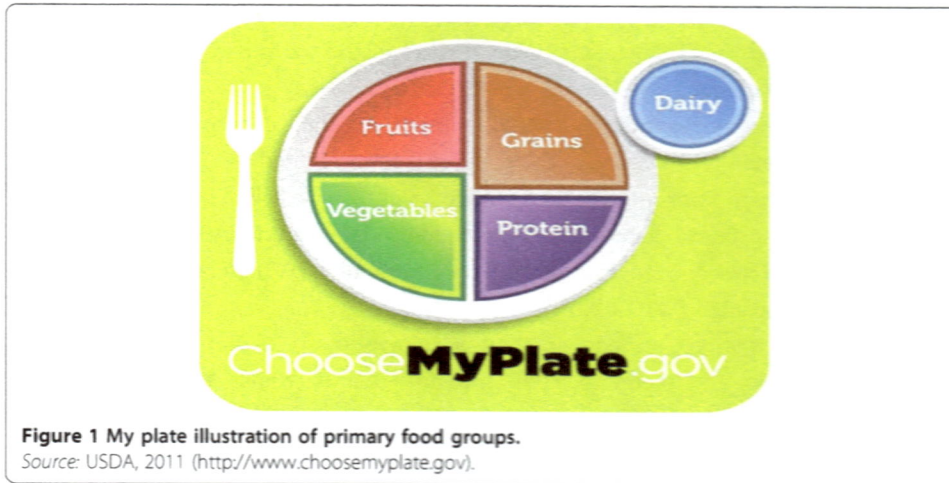

Figure 1 My plate illustration of primary food groups.
Source: USDA, 2011 (http://www.choosemyplate.gov).

satisfied by local production and by imports. Two measures commonly used in the existing literature are the self-sufficiency ratio (SSR) and the import dependency ratio (IDR). The Food and Agricultural Organization of the United Nations (FAO) defines SSR as the share of domestic production in relation to domestic food utilization, excluding stock changes, and IDR as the share of imports in relation to domestic food utilization, excluding stock changes (FAO 2001).

Several studies have utilized the SSR to analyze the structural changes of regional agricultural and food sectors. In Southeast Asia, SSRs on three categories (cereals, food, and agriculture) were employed to document the results of agricultural policy changes in Malaysia and Thailand during the 1970s and 1980s (Fitzpatrick 1991). Essentially, all three categories measured were variants of domestic commodity production divided by total domestic consumption. Malaysia saw its self-sufficiency in food increasing and self-sufficiency in cereal declining following policies to diversify its food base and displacing rice production. On the other hand, Thailand consolidated its position as a major rice exporter within a free market framework and attained higher SSRs in all three categories.

Kako (2009) defined the SSR on a calorie basis as the ratio of calorie supply from domestically produced food to the total calorie supply from all food in a country. The author found Japan's food self-sufficiency (calorie basis) decreased sharply from 79% in 1960 to 40% in 2005. The finding was attributed to a sharp appreciation of the Yen (increase purchasing power of food imports) and drastic changes in the diet of Japanese consumers.

Alternately, the SSR on a value basis is calculated as the proportion of consumer demand that is met by local production in terms of household food expenditures and farm-gate value (Holm, et al. 2000; DEFRA 2008). The former study found the overall food self-sufficiency level unchanged at 27% in the New England region in both 1975 and 1997 while the latter study found self-sufficiency for all food in the United Kingdom decreased from 76% in 1995 to 61% in 2008. Additionally, an earlier DEFRA study pointed out that since the SSR for the United Kingdom was calculated by market value, not by product weight or calorific content, it was prone to distortion, given the volatility of market prices and currency exchange rates during that time period analyzed (DEFRA 2006).

In a study of food self-sufficiency and the green revolution in India, De Janvry and Sadoulet (1991) utilized the IDR (weight basis), instead of the traditional SSR. The authors found that self-sufficiency levels for wheat, rice and coarse grains increased following the green revolution in India but without improving the nutrition requirement of its population. Kendall and Petracco (2009) defined the weighted food IDR (WIDR) as the ratio of imports to consumption in product weight for various countries in the Caribbean Basin. They found food import dependency for countries in that region (1990 to 2000) could be divided into three categories: low (WIDR<30%); medium (30%<WIDR<50%); and high (WIDR>50%).

Another study on the assessment of food sustainability in Israel provided an insightful discussion on the computed IDR measure that exceeded 100% (Gordon 2011). The author concluded that this result arises whenever exports are dependent on imports or when a certain component of an export product is imported. A relevant example here is the export of jams, which is dependent on the import of raw sugar as an ingredient. In 2008, Israel recorded IDRs for the following commodities – chicken and turkey (0.1%), beef (66%), fish (85%), bread and cereals (102%) and sweets (160%).

Both the SSR and IDR are measurable for individual food groups and aggregated groups or total. In general, we can define SSR and IDR as follows:

$$SSR = \frac{P}{P+M\text{-}X} \cdot 100\% \; (Equation \; I)$$

$$IDR = \frac{M}{P+M\text{-}X} \cdot 100\% \; (Equation \; II)$$

where P = local production of food; M = food imports; and X = food exports.

However, as shown in a later section, both measures need to be modified in order to accurately assess the extent of total food available that is satisfied by local production and by imports.

Data components and sources

In order to assess and provide a systematic estimate of food available for consumption in Hawai'i, it is necessary to first define, establish, and measure the various flows in the food supply chain. Various product flows from different sources of origin must be identified and measured at both disaggregated and aggregated levels and then grouped appropriately into a food supply matrix. Figure 2 provides a simplified illustration of this dynamic flow and food supply chain construction.

Available food from local production and imports (different group, form, and origin) are aggregated and netted out for exports (continental United States and foreign

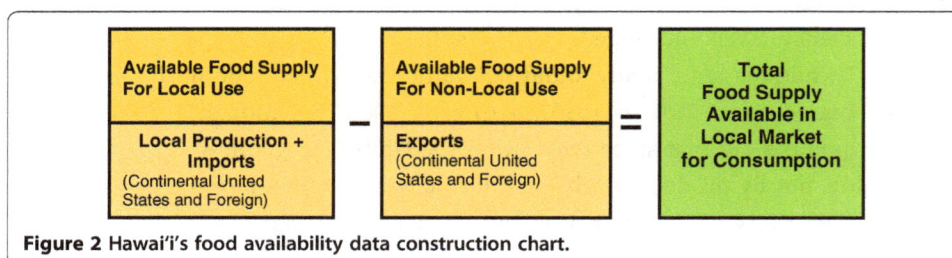

Figure 2 Hawai'i's food availability data construction chart.

countries) to arrive at the total food supply available for consumption in Hawai'i. All the necessary information is drawn from a variety of data sources, including federal and state government agencies, non-government organizations, educational institutions, cargo carriers (personal communication), wholesalers (personal communication), retailers (personal communication), industry trade groups, and independent consultants.

In this paper, the three primary sources of commercial food by origin in product weight are as follows: 1) local production data from the U.S. National Agricultural Statistics Service (NASS 2012); 2) U.S. interstate shipments data from the U.S. Army Corps of Engineers, Waterborne Commerce Statistics Center (ACE-WCSC 2012); and 3) foreign imports and exports data from the U.S. Foreign Agricultural Service (FAS 2012). Per capita consumption statistics are obtained from various sources, including the U.S. Department of Agriculture's Economic Research Service (ERS), the U.S. Census Bureau, and various Hawai'i state reports.

Results

Food supply analysis

Given Hawai'i's rapid urbanization[b] in one of the most geographically isolated areas, residents are naturally concerned about their overall food supply. Generally, this is not measurable until we can ascertain the group, type, and proportion of food consumed that is sourced locally and outside the state. This gives credence to the establishment of benchmark measures for the group (e.g., protein), type (e.g., beef), origin (e.g., continental United States), form (e.g., chilled), and quantity (e.g., weight) of food consumed in Hawai'i.

At the highest aggregated level, Figure 3 shows the food supply source and the demand destination by weight in Hawai'i, 2010. Local production accounts for just 13% of the total supply of 1.14 million tonnes of consumable food. A majority of the food sourcing (81%) is from the continental United States, while the remaining 6% is from foreign countries. On the demand side, consumption in the local market accounts for 971 thousand tonnes or 85% of the total sourced food. Exports to the continental United States stand at 14%, and the residual 1% is shipment to foreign countries.

Focusing on local commercial production, fresh fruits account for 38.9% of the total, followed by fresh vegetables at 26% and protein at 24.7%. Figure 4 shows this

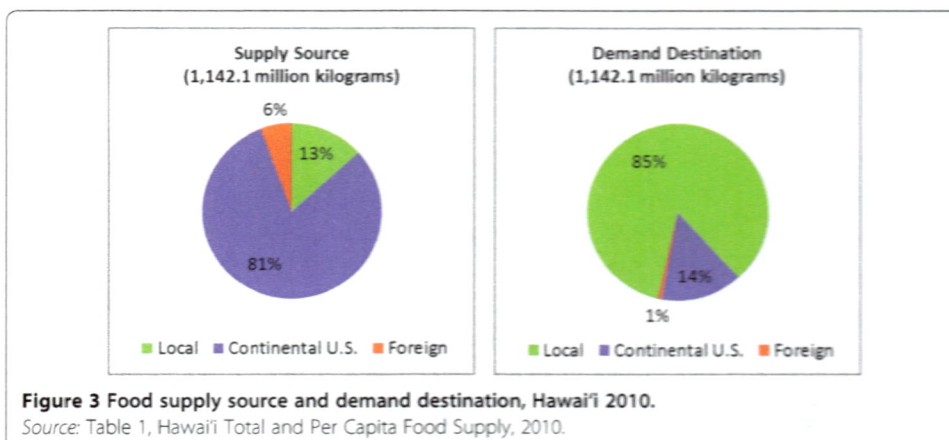

Figure 3 Food supply source and demand destination, Hawai'i 2010.
Source: Table 1, Hawai'i Total and Per Capita Food Supply, 2010.

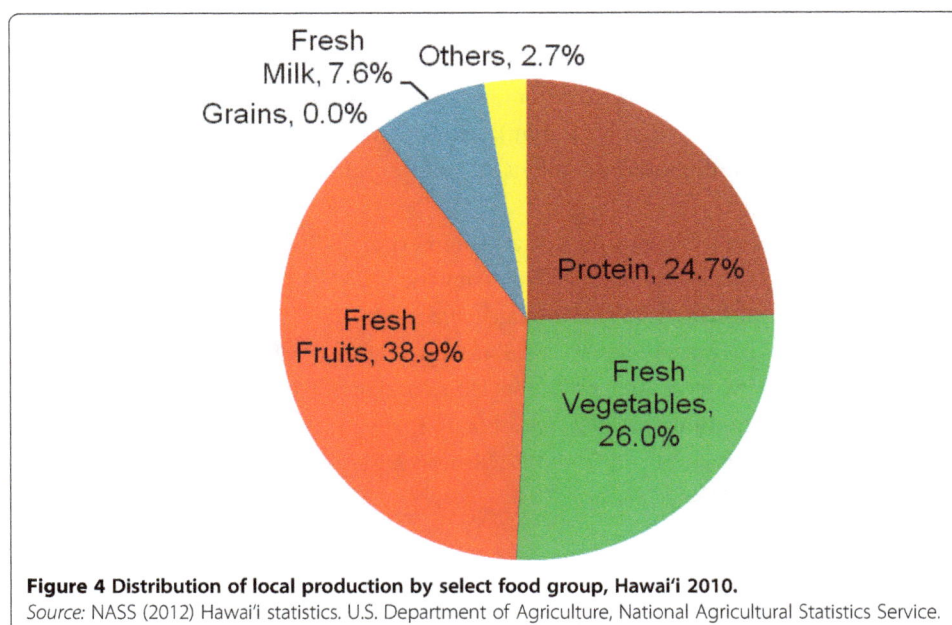

Figure 4 Distribution of local production by select food group, Hawai'i 2010.
Source: NASS (2012) Hawai'i statistics. U.S. Department of Agriculture, National Agricultural Statistics Service.

distribution by weight of the various food groups. Fresh milk accounts for 7.6% of the total, while no commercial grain production is available.

The definition of "food" applied here includes all groups (proteins, fruits, vegetables, grains, and dairy). However, due to the paucity of data as encountered in the data collection process, it was necessary to modify the food groups and to focus strictly on fresh and chilled forms of food products. Hence, food products in other forms (frozen, canned, dried, etc.) were aggregated into the residual food group "Others."

Table 1 presents the total food supply estimates for the local Hawai'i market in 2010, broken down by different food groups.[c] The core food groups include seafood (protein),

Table 1 Hawai'i total and per capita food supply[1], 2010

Food group	Local production (million kilograms)	Imports (million kilograms)		Exports (million kilograms)		Available food	
		U.S.	Foreign	U.S.	Foreign	Total (million kilograms)	Per capita[2] (kilograms)
Protein – seafood	14.4	1.2	10.8	1.5	0.3	24.6	16.7
Protein – others	23.1	97.4	5.3	23.9	0.5	101.5	69.1
Vegetables – fresh	39.5	83.9	2.5	2.2	0.0	123.7	84.2
Fruits – fresh	59.0	58.1	2.4	19.8	1.9	97.8	67.7
Grain – rice	0.0	47.0	3.3	9.2	0.0	41.0	27.9
Milk – fresh	11.4	80.9	0.0	0.0	0.0	92.4	62.9
Others	4.2	558.0	39.5	110.3	5.9	485.6	296.1
Total[3]	**151.7**	**926.7**	**63.8**	**166.9**	**8.6**	**966.6**	**657.9**

Notes:[1]Primary sources: NASS (2012) Hawai'i statistics. U.S. Department of Agriculture, National Agricultural Statistics Service; ACE-WCSC (2012) Navigation data center – domestic U.S. waterborne traffic, part 4, 2010; U.S. Army Corps of Engineers, Waterborne Commerce Statistics Center; FAS (2012) Global agricultural trade system (GATS) U.S. Department of Agriculture, Foreign Agricultural Service. [2]Based on de facto population of 1.47 million. [3]Subject to rounding errors.

other proteins, fresh vegetables, fresh fruits, rice (grain), fresh milk (dairy), and others (catch all). The estimates are expressed in the various source components (local production, imports, and exports) and measured by product weight in kilograms. Fresh fruits, fresh vegetables, and other proteins are the three largest food groups sourced from local production. Other proteins, fresh vegetables, and fresh milk are the largest import components. Combined, the leading available food groups in the Hawai'i market are fresh vegetables, other proteins, fresh fruits, and fresh milk.

On a per capita food measurement basis, fresh vegetables again lead with 84.2 kilograms, followed by other proteins at 69.1 kilograms, fresh fruits at 67.7 kilograms, fresh milk at 62.9 kilograms, and rice at 27.9 kilograms. Overall, we estimate the available food per capita for Hawai'i at 657.9 kilograms in 2010. With a reported $3.68 billion (2004–2005 dollars) spent on food annually in Hawai'i, this translates into an inflation adjusted estimate of $4.52 billion[d] spent on food in 2010. The average cost of food available in the local market is $4.66 per kilogram ($4.5 billion/966.6 thousand tonnes).

Self-sufficiency ratio and import dependency ratio

Figure 5 shows the SSR of defined food groups and the overall total for Hawai'i. It shows that Hawai'i has an overall food SSR of 15.7%, and that fresh fruits have the highest level of self-sufficiency among the food groups at 60.4%. Seafood protein follows next at 58.7% and fresh vegetables at 31.9%. No rice is produced commercially in Hawai'i. Finally, the SSR of fresh milk in Hawai'i stands at 12.4%.

Likewise, Figure 6 shows the IDR of defined food groups and the overall total for Hawai'i in 2010. Hawai'i has an overall food IDR of 102.5%, and rice has the highest level of import dependency for a food group, at 122.5%. Other proteins follow next, at 101.3%, and other food groups, such as seafood, fresh fruits, fresh vegetables, and fresh milk range from 48.7% to 87.6%. While the IDR exceeding 100% is intuitively confusing and appears to be a measurement error, this is not the case here. Rather, it indicates the existence of food imports into Hawai'i that are then turned around and re-exported to other markets. Entreport markets, such as Singapore or Israel, consistently exhibited IDR significantly higher than 100% (Mikic and Gilbert 2007; Gordon 2011). The measures exceeding 100% are a reflection of re-exports embedded in the raw data collected from official sources. Hence, they do not accurately

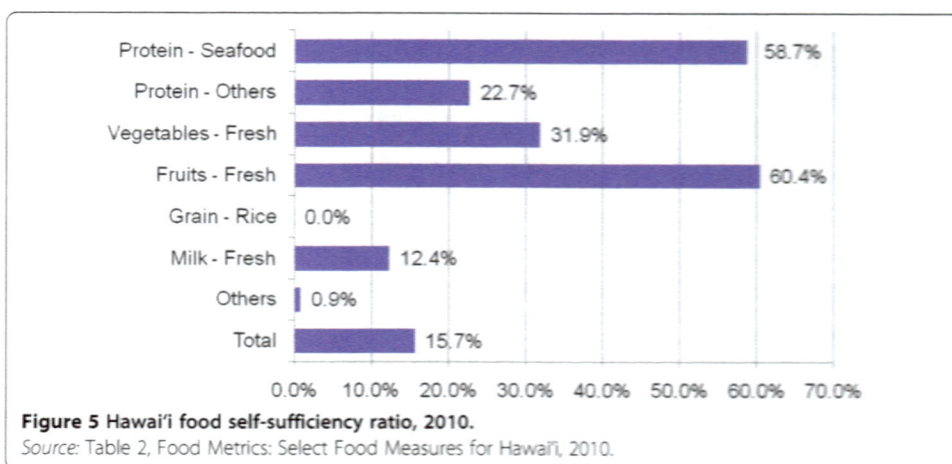

Figure 5 Hawai'i food self-sufficiency ratio, 2010.
Source: Table 2, Food Metrics: Select Food Measures for Hawai'i, 2010.

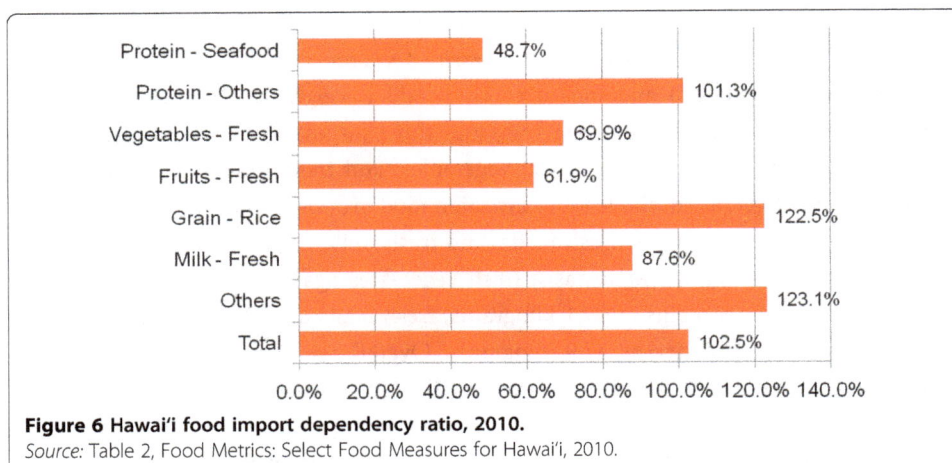

Figure 6 Hawai'i food import dependency ratio, 2010.
Source: Table 2, Food Metrics: Select Food Measures for Hawai'i, 2010.

reflect food imports for consumption in Hawai'i. Likewise, SSR is clearly not the complement of IDR, as the two figures do not sum to 100%.

A modified set of measurements

In the previous section, the IDR exceeded 100%. It is clear this measure is biased upwards when re-exports are not accounted for. Likewise, the SSR is biased upwards when exports are not accounted for in the defined food groups. Intuitively, the SSR serves as a better measure of *potential* local production to satisfy the net demand or local consumption of food. To derive more definitive measures to assess food self-sufficiency and import dependency in Hawai'i, we propose to modify the SSR and IDR as follows:

Define Xm = re-exports of food imports; and Xp = food exports from local production.

$$SSR' = \frac{P}{P+M\text{-}Xm\text{-}Xp} \cdot 100\% \ (Equation \ I')$$

$$IDR' = \frac{M}{P+M\text{-}Xm\text{-}Xp} \cdot 100\% \ (Equation \ II')$$

If we take into account the amount of exports from local production, we can redefine Equation I' as the Localization Ratio (LR):

$$LR = \frac{P\text{-}Xp}{P+M\text{-}Xm\text{-}Xp} \cdot 100\% \ (Equation \ III)$$

Similarly, we can redefine Equation II' as the Modified Import Dependency Ratio (MIDR):

$$MIDR = \frac{M\text{-}Xm}{P+M\text{-}Xm\text{-}Xp} \cdot 100\% \ (Equation \ IV)$$

The sum of LR and MIDR (Equation III and Equation IV) is now 100%:

$$LR+MIDR = \frac{P\text{-}Xp+M\text{-}Xm}{P+M\text{-}Xm\text{-}Xp} \cdot 100\% = 100\% \ (Equation \ V)$$

Table 2 shows the SSR, IDR, LR and MIDR of defined food groups for Hawai'i in 2010. It is worth noting the LR is lower than the SSR in each food group, except for rice, fresh milk and others, which remained unchanged, due to the absence of exports from local production. The variation between LR and SSR is progressively larger in food groups with a higher level of exports from local production, e.g. fresh fruits. Likewise, the IDR is biased upwards compared to the MIDR of equivalent food groups, whenever re-exports are not accounted for in the raw dataset. Overall, LR is a more accurate measure on the extent of local food consumption that is sourced from local production, in the absence of stock changes. The total LR estimate indicates only 11.6% of available food for consumption in Hawai'i was sourced from local production in 2010. Likewise, the MIDR estimate shows 88.4% of available food in Hawai'i was sourced from imports.

Findings by Major Food Group

1. Seafood protein: Seafood is a significant component in the food diet of residents in Hawai'i. On a per capita basis, commercial seafood consumption is estimated at 12.9 kilograms for Hawai'i, or 1.8 times more than the 7.2 kilograms for the overall United States. This measure increases to 16.7 kilograms when non-commercial (recreational) catch is included. The average equivalent commercial measure for Hawai'i in the 1970s was 9.5 kilograms (Hudgins 1980), about 3.4 kilograms less than current per capita seafood consumption. In general, Hawai'i consumes more fresh and frozen finfish (yellowfin tuna, bigeye tuna and salmon), whereas the rest of the country consumes more shellfish and processed seafood (shrimp and canned tuna).

   Seafood supplies are sourced primarily from local and foreign imports. Collectively, local landings, aquaculture, and noncommercial catch make up 51% of total available seafood supply in Hawai'i. Foreign imports account for 44%, and imports from the continental United States fill the remaining 5% (Loke et al. 2012). According to the U.S. Foreign Agricultural Service (FAS), the leading direct foreign sources of seafood imports by weight were from Taiwan, Japan, New Zealand, the Philippines, and the Marshall Islands.

2. Other proteins: Products in this group include red meat (beef, veal, pork, and lamb), poultry (chicken, duck, and turkey), and nuts that are produced locally and imported. In 2010, net supplies from all sources totaled 101 thousand tonnes for this food group. Local production is 9.3% of total market requirement, and the state

Table 2 Food matrix: select food measures for Hawai'i, 2010

	Group	SSR	IDR	LR	MIDR
1	Protein – seafood	58.7%	48.7%	51.3%	48.7%
2	Protein – others	22.7%	101.3%	9.3%	90.7%
3	Vegetables – fresh	31.9%	69.9%	30.1%	69.9%
4	Fruits – fresh	60.4%	61.9%	38.1%	61.9%
5	Grain – rice	0.0%	122.5%	0.0%	100.0%
6	Milk – fresh	12.4%	87.6%	12.4%	87.6%
7	Others	0.9%	123.1%	0.9%	99.1%
	Total	15.7%	102.5%	11.6%	88.4%

Sources: Table 1, Hawai'i Total and Per Capita Food Supply, 2010; NASS (2012) Hawai'i statistics. U.S. Department of Agriculture, National Agricultural Statistics Service; and personal communication with local wholesalers and industry analysts.

is dependent on imports for the remaining 90.7%, mainly from the continental United States. The Hawai'i market supply level in 2010 is about 37% higher than the 73.9 thousand tonnes in 1980.

Per capita consumption of other proteins (excluding nuts) is estimated at 65.9 kilograms, 19% lower than the overall United States measure of 78.5 kilograms. This estimate in 2010 is also 4.5% lower than the per capita estimate of 68.9 kilograms in 1980. The trend in per capita consumption of other proteins (meat) in Hawai'i appears declining slowly over time as consumers switch to healthier, alternate substitutes.

When commercial seafood protein is combined with other proteins (excluding nuts), the combined per capita consumption is 79.1 kilograms in Hawai'i, only 8% lower than the overall United States measure of 85.7 kilograms. This combined estimate in 2010 is marginally lower (0.9%) compared to the per capita estimate of 78.4 kilograms in 1980.

3. Fresh Vegetables: Products in this group include leafy and non-leafy greens, sweet corn, tubers (ginger root, potatoes, and taro), and specialty greens that are both grown locally and imported. Net fresh vegetable supplies from all sources totaled 124 thousand tonnes in 2010. Local production is 30% of total market requirement, and the state is dependent on imports for the remaining 70%, mainly from the continental United States. The market supply in 2010 is about 56% higher than the 79.4 thousand tonnes recorded in 1980.

   Hawai'i's per capita consumption of fresh vegetables is estimated at 84.2 kilograms in 2010, slightly less than the overall United States measure of 84.9 kilograms. In contrast, this estimate is 12% higher than the per capita estimate of 75.3 kilograms in 1980. The rising trend in per capita consumption of fresh vegetables is likely to continue over time as more residents strive to consume the recommended five servings of fresh fruits and vegetables (FFVs) on a daily basis to increase fiber intake and to realize a healthier diet in their personal lives.

4. Fresh Fruits: Products in this group include tropical fruits grown locally (bananas, guavas, papayas, pineapples, watermelons, etc.) and imported fruits (apples, citrus fruits, berries, stone fruits, etc.). Net fresh fruit supplies from all sources totaled 98 thousand tonnes in 2010. Local production is 38% of total market requirement, and the state is dependent on imports for the remaining 62%, mainly from the continental United States. The market supply in 2010 is about 2.6 times the equivalent measure of 36.9 thousand tonnes recorded in 1980.

   It is worth noting that 37% of Hawai'i's estimated fruit production of 59 thousand tonnes in 2010 is exported. Should we choose to redirect fresh fruit exports to the local market, we could satisfy 60% of total consumption requirement.

   The per capita consumption of fresh fruits is estimated at 67.7 kilograms, moderately higher (17%) compared to the overall United States measure of 58 kilograms. This estimate in 2010 is close to double the per capita estimate of 35 kilograms recorded in 1980. Similar to fresh vegetables, the rising trend in per capita consumption of fresh fruits is likely to continue over time as more residents strive to consume five servings of fresh fruits and vegetables (FFVs) each day. Additionally, the rapid growth of tourism in Hawai'i since 1980 has necessitated the increased provision of non-tropical fruits that satisfy the taste preferences of visitors. In that same time period, the average number of visitors present per day in Hawai'i, increased 85%, from 96,406 visitors in 1980 to 177,949 visitors in 2010[e].

5. Fresh Milk: As recently as the early 1980s, Hawai'i produced all fresh milk (dairy) that was consumed in the state. This was a startling achievement, considering only 24.4% market supply was sourced locally in the 1930s. Since then, a host of less favorable economic circumstances has turned against the industry, wiping out all commercial dairy farms on the island of Oahu, and leaving only two on the island of Hawai'i. In 2010, local production supplied 12.4% of total fresh milk available in the local market. In other words, Hawai'i is 87.6% dependent on fresh milk imports from outside sources (continental United States).

In 2010, per capita consumption of fresh milk in Hawai'i is estimated at 62.9 kilograms. This is significantly lower than the per capita overall United States consumption of 92.4 kilograms. In 1980, the same measure for Hawai'i was 74 kilograms, again lower than the comparable national measure then of 111.4 kilograms. The proportion of per capita fresh milk consumption between Hawai'i and the United States was 0.68 in 2010 and 0.665 in 1980. This difference between Hawai'i and the national measure is historical, and due in large part to the population mix in Hawai'i. There is a proportionately large population of residents of Asian descent, many of whom are lactose intolerant, which contributes to a lower propensity to consume fresh milk.

6. Rice: In the conclusion of his 1937 publication, H.H. Warner identified rice as the single most important food item that Hawai'i imports from the outside world. During that period in history, the Territory of Hawai'i produced only 4.1% of its total rice requirement, and was increasingly threatened by lower cost, mechanized producers in California. Per capita consumption of rice in Hawai'i then was reportedly 40 times higher than in the United States.

Today, there is no known commercial production of rice in Hawai'i. We are totally dependent on imports, particularly from the continental United States. About 6% of the total requirement is foreign imports, mainly specialty rice from Thailand. Over time, per capita consumption of rice in Hawai'i declined to an estimated 27.9 kilograms in 2010 from 34.3 kilograms in the mid-1970s[f]. This measure is now only three times more than the 9.6 kilograms per capita consumption in the overall United States.

Discussion

On a per capita basis, Hawai'i consumes more fresh fruits, rice, and seafood as compared to average consumers in the United States. In contrast, residents in Hawai'i consume less than average United States' residents in food groups such as fresh or chilled meats, fresh milk, and fresh vegetables (marginally less in the latter group). It is plausible that lower consumption of fresh or chilled meats is offset by a higher consumption of canned or processed meats. Anecdotal evidence suggests SPAM®(luncheon meat) is a local favorite, and Hawai'i has been cited often as the SPAM® capital of the world for its high per capita consumption. Fresh milk consumption has been historically lower as many residents in the local community avoid lactose in dairy milk.

With the 2010 measures currently available, it becomes possible to compare the extent of local consumption arising from local production over time (historical analysis). One of the earliest and best documented studies on food sourcing in Hawai'i was conducted by the Agricultural Extension Service, University of Hawai'i in 1937. This study was conducted with the primary objective of documenting the effects of the Hawai'i maritime strike on food supply in 1936–1937. It found the overall food consumption sourced locally in Hawai'i was 37.1% as measured by product weight between January 1934 to October 1936. The corresponding available food per capita then was 518.3 kilograms, some 27% lower than the comparable 2010 per capita estimate of 657.9 kilograms.

Likewise, a state Department of Agriculture planning document[g] provided equivalent measures for some commodity food groups in 1980. Figure 7 presents a historical comparison of the proportion of food sourced locally by core food groups in Hawai'i in 1934–36, 1980, and 2010. Ironically, all years reviewed were preceded by a period of tumultuous economic downturn in the United States. The period, 1934–36, marked the official recovery following the *Great Depression* (August 1929 to March 1933); the first seven months of 1980 saw an enduring Organization of Petroleum Exporting Countries (OPEC) induced recession (January 1980 to July 1980); and 2010 marked the economic recovery in the aftermath of the *Great Recession* (December 2007 to June 2009)[h].

While the proportion of overall food sourced locally declined precipitously from 37.1% in 1934–36 to 11.6% in 2010, falling 25.5% during the intervening 74-year period, the reductions in corresponding core food groups were generally, less dramatic. The two exceptions noted were other proteins and fresh vegetables which dropped 35% and 30% respectively. In sharp contrast to the prevailing trend, seafood sourced locally increased by 2.2% in that same time period.

Lessons learned

As public discourse on available food for consumption and local production continues in Hawai'i, it becomes apparent that state benchmark estimates are required to track market requirements and their supply sources. Food import measures are not readily available and this

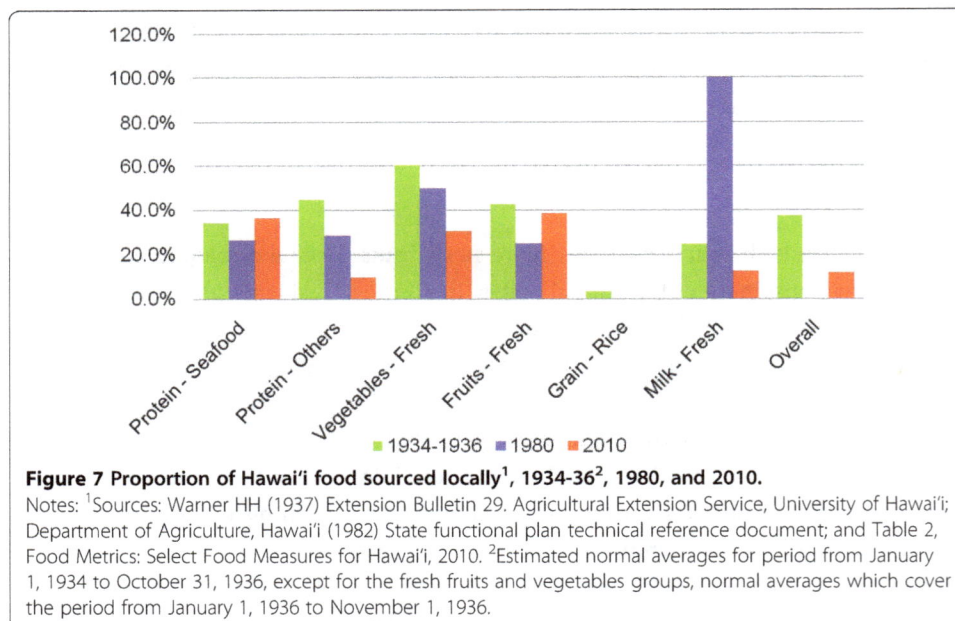

Figure 7 Proportion of Hawai'i food sourced locally[1], 1934-36[2], 1980, and 2010.
Notes: [1]Sources: Warner HH (1937) Extension Bulletin 29. Agricultural Extension Service, University of Hawai'i; Department of Agriculture, Hawai'i (1982) State functional plan technical reference document; and Table 2, Food Metrics: Select Food Measures for Hawai'i, 2010. [2]Estimated normal averages for period from January 1, 1934 to October 31, 1936, except for the fresh fruits and vegetables groups, normal averages which cover the period from January 1, 1936 to November 1, 1936.

paper is intended to fill that gap by providing a more informed assessment of the market place by bringing together information from published data sources, industry technical reports, and informed industry sources. While the goal is straight forward, the technical efforts required to benchmark the estimates from diverging sources into a meaningful common denominator are far more challenging. The discussion below outlines some of the technical difficulties encountered in this benchmarking effort.

Figure 2 illustrates the blueprint for constructing the food consumption and supply chain flows in Hawai'i. With the methodology established, we proceeded to define food groups and sub-groups; select food products for inclusion; and to collate relevant information from core datasets (local production, imports, and exports). The separate procedures and related technical difficulties encountered are expressed as follows:

1. Classification: This initial step is to establish food groups and then break these into finer subgroups whenever possible. Ideally, the groups selected should be defined in the applicable datasets relating to local production, as well as imports and exports (domestic and foreign). A crucial step here is to determine the level of aggregation or disaggregation of food products. An equally important step is to determine the food products to be included or excluded. In this paper, we adopted the major food groups as outlined in the U.S. Department of Agriculture (USDA) My Plate concept – dairy, grains, protein, fruits, and vegetables (Figure 1). Subgroups are also designated, such as for protein – seafood, beef, veal, pork, other meats, and nuts; and fruits – fresh, frozen, dried and prepared. To facilitate meaningful comparisons with other established metrics, we excluded beverages (coffee, tea, water, malt extracts, wine, spirits, liqueur, beer, etc.), seeds and spores, and live animals.

2. Standardization: Once classification is complete, the task of compiling, aggregating and converting the many food products with different unit measures into a common unit measure begins. Various databases and data sources utilize different measuring units. American databases such as NASS, adopt measuring units ranging from pounds, short tons, gallons, and actual units (number of eggs or heads of cattle). At the local production level, there are various product settings such as harvested, utilized, processed or dressed. International databases (foreign imports and exports) adopt the metric system and report measures ranging from kilograms, metric tons (tonnes), liters, kiloliters, and actual units.

   Furthermore, the conversion of volume measure to weight measure, for instance kiloliters to pounds or gallons to kilograms, requires knowledge of the specific liquid's density. For example, one liter of water (at 4 degrees Celsius) is about one kilogram and converts to about 2.2 pounds, whereas one liter of olive oil is about 0.92 kilogram and converts to about 2.02 pounds. Finally, it is important to convert the measurement of food items to their most consumable form. For example, livestock products are defined in dressed weight as opposed to live weight and seafood is defined in edible weights as opposed to product weight. Overall, this procedure can be competently accomplished with more resources, particularly, with relevant datasets.

3. Reconfiguration: When specific items from various datasets or data sources are not available or not clearly delineated, it is necessary to reconfigure the initially defined food groups and subgroups. In this study, we were not able to differentiate canned, dried or processed, fruits and vegetables. Hence, we reconfigured fruits and vegetables to fresh

fruits and fresh vegetables. The same challenge confronts grains and dairy. Hence, we redefined them as rice (grains) and fresh milk (dairy). All the undifferentiated products were aggregated into the residual (catch all) food group "Others." By reconfiguring the food groups/subgroups, we can provide meaningful comparative metrics to the overall benchmarking process.

4. Validation: This crucial procedure assesses how significant and relevant the various compiled statistics and estimates are in the various food groups. The rationality and consistency tests within and across time frames are important to establish the validity of the information presented. Various challenges that lurk around the corner include double counting and benchmarking against inappropriate/misleading industry metrics, including metrics that are ill defined.

Double counting is a real hazard when reconfiguring the various foodgroups/subgroups, redistributing the weight measure of various food products, and in measuring processed (value-added) products. The same challenge is encountered when aggregating out-shipment (export) volumes from various island ports. Summing up volumes from each island port will lead to double counting since a large volume of exports are shipped to Honolulu before being shipped to the continental United States. Secondly, out-shipment from Honolulu, can also imply in-shipment to neighbor island ports.

Inappropriate/misleading industry metrics can arise when specific local commodity measures are estimated by utilizing per capita national measures (as opposed to state or local) as proxies. This practice skews the estimate of total commodity production away from its true value. The unique composition of Hawai'i's population does not always lends itself to similar consumption patterns of residents in other American states. We have established that Hawai'i residents consume more seafood (1.8 times), more rice (3 times) but less fresh milk (1/3 times) when compared to all United States residents on a per capita basis. Finally, only comparable measures with similar definitions should be utilized for validating estimated measures. In this paper, we included nuts in the protein food group and this food item is retracted when comparing red meat and poultry, the standard for defining protein at the national level by USDA.

While we make estimates at different levels, it is assumed the databases or data sources are credible and reasonably accurate, at least at the higher (aggregated) level. Hence estimates at the higher level are more robust than those at the lower (disaggregated) level. Plainly, the sum of all food groups across each dataset or data source is more complete than the sum of food subgroups or products that may be missing in one or more dataset or data source.

Conclusions

This paper establishes a set of benchmark estimates on food available for market consumption and its supply sources in Hawai'i. Overall, we estimated the total food available at 966.6 million kilograms, or a per capita consumption of 657.9 kilograms in 2010. Fresh vegetables, other proteins, and fresh fruits were the leading defined food groups. The vast majority of this available food is imported from the continental United States (81%), with 6% from foreign countries.

In assessing the extent of food localization in Hawai'i, we adopted the SSR and its complementary measure, the IDR. Hawai'i had an overall food SSR of 15.7% and an overall food IDR of 102.5% in 2010. Obviously, the two complementary measures do not sum up to 100%. This result led us to conclude that SSR was a better measure of "potential" local production to satisfy local market consumption of food in Hawai'i. Likewise, the IDR exceeding 100% indicates the existence of food imports that are then turned around and re-exported to other markets.

To provide a more accurate assessment of food localization, we suggested two modified measures of self-sufficiency and import dependency, namely the localization ratio (LR) and the modified import dependency ratio (MIDR). The overall LR estimate reveals that only 11.6% of available food for consumption in Hawai'i was sourced from local production in 2010. Likewise, the MIDR estimate shows 88.4% of available food in Hawai'i was sourced from imports. Beyond that, we have also established that the consumption patterns of some food groups are quite different for residents in Hawai'i than for those in the overall United States.

As a final point, a historical comparison of total food consumption suggests that Hawai'i's per capita measure increased to 657.9 kilograms in 2010, as compared to only 518.3 kilograms in 1936. However, Hawai'i had a lower proportion of food sourced locally in 2010, with a localization ratio (LR) of 11.6%; this is much lower compared to the ratio of 37.1% in 1936. Within defined food groups, Hawai'i had a higher LR in seafood in 2010 than in 1936. Lower LRs were recorded for the remaining food groups defined.

The analytical framework presented in this paper can be applied to other small, open (island or regional) economies with a food localization agenda, as it provides a more discrete and appropriate set of measurements, as well as offering the lessons gained through Hawai'i's experience and challenges in the data-collating process. Additionally, the procedures and food groups defined are particularly applicable to island economies, which are less resource endowed in the production of grains and are far too often neglected in mainstream studies on food self-sufficiency.

With the metrics and benchmark estimates now realized, there are potential implications for food policy decisions in Hawai'i. Since the SSR and LR rely on both local production (supply) and market consumption of food (demand), it may not be optimal for the state to establish an arbitrary annual target measure for food localization. First, agricultural production fluctuates substantially from year to year, usually constrained by periods of drought, heavy rain and floods, and other *force majeure* events. Likewise, food consumption is subject to fluctuations over time, usually influenced by changes in consumer preferences and income. Secondly, it may enhance the state's welfare to support the cultivation of certain food groups for which Hawai'i has a comparative advantage in production and taste preference and which are substitutes for products that require a large volume of imports. One clear example is seafood, which can be substituted for other protein products. At a different level of self-sufficiency, it may be advantageous to support the exports of certain food groups and not divert them into local markets. Clearly, earnings from the export of tropical fruits (pineapples, papayas, avocados) can be utilized to fund the imports of temperate-climate fruits such as apples, citrus, and stone fruits. Likewise, the export of sweet potatoes, fish, and basil can be exchanged for rice imports. Agricultural export earnings can also be utilized to purchase needed factor inputs such as fertilizers, pesticides, packaging materials, and farm implements.

Finally, the logical extension for continuing research in Hawai'i's food supply and consumption matrix is to expand the defined food groups and subgroups and to further delineate them into food forms such as dried, canned, or processed. Furthermore, it is equally appropriate to assess the various food groups and subgroups that have the greatest potential to achieve a higher degree of food localization in Hawai'i. The additional knowledge arising from such an assessment could further alleviate many of the concerns expressed by Hawai'i residents as to the vulnerability of the state's food supply chain in the midst of continuing volatility in global markets.

Endnotes

[a]Warner HH (1937) *Hawai'i's food supply and the maritime strike of 1936-37*. Extension Bulletin 29. Agricultural Extension Service, University of Hawai'i.

[b]Hawaii's rapid urbanization is well reflected in Honolulu's growing traffic congestion. See Honolulu StarAdvertiser (Apr 04, 2013) Honolulu third-worst for traffic congestion. http://www.staradvertiser.com/news/breaking/20130404_Honolulu_thirdworst_for_traffic_congestion.html?id=201473771 Accessed 9 Apr 2013.

[c]All data sources utilized in this table are reported in quantities (usually weight measures available) and not in other metrics (e.g., dollars, calories or nutrition values). The weights used consistently here are as reported and have not undergone conversions, except for eggs and poultry (unit count to product weight and live to dressed weight). Seafood is reported in edible weight as sourced verbatim from a previous study (conversions from product to edible weight).

[d]This estimate is derived from CPI-U Honolulu (Food and Beverages), Table 14.4 (2004, 2005, and 2010), State of Hawai'i Databook from a base value of $3.68 billion, sourced from the BLS Consumer Expenditures Survey, 2004-2005.

[e]See 2011 State of Hawai'i Databook. Table 7.03, Visitors arrival and average daily census: 1966 to 2011.

[f]This is according to Lee and Bittenbender (2008) *Agriculture*. Paper for Hawai'i sustainability 2050. College of Tropical Agriculture and Human Resources, University of Hawai'i at Manoa.

[g]DOA (1982) *A state functional plan technical reference document*. Department of Agriculture, Hawai'i.

[h]Respective timelines are sourced from the National Bureau of Economic Research (NBER) at http://www.nber.org/cycles.html Accessed 12 Dec 2012.

Competing interests
The authors declare that they have no competing interests.

Author's contributions
MKL carried the study plan, collected data, generated quantitative results, and drafted the manuscript, figures, and tables. PSL helped to develop the methodology, analyzed the results, and reviewed the manuscript. Both authors read and approved the final manuscript.

Acknowledgements
This paper benefited greatly from the constructive comments and suggestions of two anonymous journal referees, Dr. James Mak, Emeritus Professor of Economics, UHM and Frederika Bain, CTAHR Office of Communication Services, UHM. Kristopher Keahiolalo, Graduate Assistant, NREM-UHM provided able research support. Responsibility for the final content rests with the authors. The authors gratefully acknowledge the funding support of the Hawai'i Department of Agriculture, the College of Tropical Agriculture and Human Resources, UHM, under the Research Supplemental Funds Program (Award No. HAW01122-H), and the U.S. Department of Agriculture, Agricultural Research Service, under the Specific Cooperative Agreement "Agricultural Post- Harvest, Value Added Products and Processing Program" (Award

No. 58-5320-7-664). Finally, we acknowledge invaluable assistance from numerous industry contacts in sharing information pertinent to this project.

References

ACE-WCSC (2012) Navigation data center – domestic U.S. waterborne traffic, part 4, 2010. U.S. Army Corps of Engineers. Waterborne Commerce Statistics Center, New Orleans, LA. http://www.navigationdatacenter.us/wcsc/webpub10/webpubpart-4.htm. Accessed 16 Jul 2012

De Janvry A, Sadoulet E (1991) Food self-sufficiency and food security in India: Achievements and contradictions. In: Ruppel FJ, Kellogg ED (ed) National and regional self-sufficiency goals. Lynne Rienner, Boulder & London

DEFRA (2006) Food security and the UK: An evidence and analysis paper. UK. Department for Environment, Food and Rural Affairs, London. http://archive.defra.gov.uk/evidence/economics/foodfarm/reports/documents/foodsecurity.pdf. Accessed 5 Apr 2013

DEFRA (2008) Food statistics pocket book 2008. U.K. Department for Environment, Food and Rural Affairs, London. http://webarchive.nationalarchives.gov.uk/20130123162956/http://www.defra.gov.uk/statistics/files/defra-stats-foodfarm-food-pocketbook-2008.pdf. Accessed 5 Apr 2013

FAO (2001) Food balance sheets – A handbook. Food and Agricultural Organization of the United Nations, Rome. http://www.fao.org/docrep/003/x9892e/x9892e00.htm. Accessed 26 Oct 2012

FAS (2012) Global agricultural trade system (GATS) U.S. Department of Agriculture, Foreign Agricultural Service, Washington D.C. http://www.fas.usda.gov/gats/default.aspx. Accessed 12 Jun 2012

Fitzpatrick E (1991) Agricultural self-sufficiency in Southeast Asia: Malaysia and Thailand. In: Ruppel FJ, Kellogg ED (ed) National and regional self-sufficiency goals. Lynne Rienner, Boulder & London.

Gordon U (2011) Program for an assessment of food sustainability in Israel. Arava Institute for Environmental Studies, Ketura, Israel. http://www.arava.org/userfiles/file/Director/_Food.doc. Accessed 5 Apr 2013

Halweil B (2004) Eat here: Reclaiming homegrown pleasures in a global supermarket. W.W. Norton & Co., New York

Heady D, Fan S (2008) Anatomy of a crisis: the causes and consequences of surging food prices. Agricultural Economics. 39:375–391

Holm D, Rogers R, Lass D (2000) Food self-sufficiency in the New England states, 1975–1997. Department of Resource Economics, University of Massachusetts, Amherst, MA. http://www.massbenchmarks.org/publications/studies/pdf/foodself00.pdf. Accessed 26 Oct 2012

Hudgins L (1980) Per capita annual utilization and consumption of fish and shellfish in Hawai'i, 1970–77. Marine Fisheries Review. 42:16–20

Kako T (2009) Sharp decline in the food self-sufficiency ratio in Japan and its future prospects. Paper presented at the International Association of Agricultural Economists Conference, Beijing, China. 16–22 August 2009

Kendall P, Petracco M (2009) The current state and future of Caribbean agriculture. Journal of Sustainable Agriculture. 33(7):780-797

Leung PS, Loke M (2008) Economic impacts of improving Hawai'i's food self-sufficiency. Economic Issues EI-16 (December 2008). College of Tropical Agriculture and Human Resources, University of Hawai'i at Manoa, Honolulu, HI. http://www.ctahr.hawaii.edu/oc/freepubs/pdf/EI-16.pdf. Accessed 12 Dec 2012

Loke M, Geslani C, Takenaka B, Leung PS (2012) Seafood consumption and supply sources in Hawai'i, 2000–2009. Marine Fisheries Review 74(4):44–51

Melrose J, Delparte D (2012) Hawai'i county food self-sufficiency baseline 2012. University of Hawai'i, Hilo, HI. http://geodata.sdal.hilo.hawaii.edu/GEODATA/COH_Ag_Project.html. Accessed 26 Oct 2012

Mikic M, Gilbert J (2007) Trade statistics in policymaking: A handbook of commonly used trade indices and indicators. Economic and Social Commission for Asia and the Pacific (ESCAP), Bangkok, Thailand. http://www.unescap.org/tid/aptiad/Handbook2.pdf. Accessed 11 Jan 2013

NASS (2012) Hawai'i statistics. U.S. Department of Agriculture, National Agricultural Statistics Service, Washington D.C. http://www.nass.usda.gov/Statistics_by_State/Hawaii/index.asp. Accessed 2 Jul 2012

OP-DBEDT (2012) Increased food security and food self-sufficiency strategy. Office of Planning, Department of Business, Economic Development and Tourism, Hawai'i. http://files.hawaii.gov/dbedt/op/spb/INCREASED_FOOD_SECURITY_AND_FOOD_SELF_SUFFICIENCY_STRATEGY.pdf. Accessed 12 Dec 2012

Page C, Bony L, Schewel L (2007) Island of Hawai'i whole system project phase I report. Rocky Mountain Institute, Boulder, CO. http://www.kohalacenter.org/pdf/hi_wsp_2.pdf Accessed 26 Oct 2012

Ulupono I (2011) Local food market demand study of O'ahu shoppers: Executive summary. https://dl.dropboxusercontent.com/u/40878762/Local%20Food%20Market%20Demand%20Study%20Executive%20Summary.pdf. Accessed 12 Dec 2012

Horner, J., R. Milhollin, and W. Prewitt (2012) Economics of Pasture-Based Dairies. Publication m192, University of Missouri Extension Service, Columbai, MO

Khanal AR, Gillespie J (2013) Adoption and productivity of breeding technologies: evidence from U.S. dairy farms. AgBioForum 16(1):53–65

Khanal A, Gillespie J, MacDonald J (2010) Adoption of technology, management practices, and production systems in U.S. milk production. J. Dairy Sci 93(12):6012–6022

Mayen C, Balagtas J, Alexander C (2010) Technology adoption and technical efficiency: organic and conventional dairy farms in the United States. Am J Agric Econ 92(1):181–195

McBride W, Greene C (2009) Costs of organic milk production on U.S. dairy farms. Rev Agric Econ 31(4):793–813

McBride W, Short S, El-Osta H (2004) The adoption and impact of bovine somatotropin on U.S. dairy farms. Rev Agric Econ 26:472–488

Mishra A, Perry J (1999) Forward contracting of inputs: A farm-level analysis. J Agribus 17(2):77–91

National Research Council, Panel to Review USDA's Agricultural Resource Management Survey (2008) Understanding American Agriculture: Challenges for the Agricultural Resource Management Survey. National Academy Press, Washington, DC

Pesaran MH, Shin Y, Smith RP (1999) Pooled mean group estimation of dynamic heterogeneous panels. J Am Stat Assoc 94(446):621–634

Pruitt R, Gillespie J, Nehring R, Qushim B (2012) Adoption of technology, management practices, and production systems by U.S. beef cow-calf producers. J Agric Appl Econ 44(2):203–222

Short S (2000) Structure, management, and performance characteristics of U.S. dairy farms. Agricultural Handbook No. 720, USDA – Economic Research Service, September, Washington, DC

Tauer LW (1998) Cost of production for stanchion versus parlor milking in New York. J Dairy Sci 81(2):567–569

Tauer LW (2009) Estimation of treatment effects of recombinant bovine somatotropin using matching samples. Rev Agric Econ 31(3):411–423

Turner L, Skele T (2007) Dairy herd batch calving: Findings from the sustainable dairy farm systems for profit project. M5 Info Series – 133 Batch Calving, Department of Primary Industries and Fisheries, Queensland, Australia

Ward CE, Vestal MK, Doye DG, Lalman DL (2008) Factors affecting adoption of cow- calf production practices in Oklahoma. J Agric Appl Econ 40(3):851–63

Wiegel KA (2004) "Exploring the role of sexed semen in dairy production systems. J Dairy Sci 87(E-Suppl):E120–E130

Zar JH (1984) Biostatistical Analysis, 2nd edn. Prentice-Hall, Englewood Cliffs, NJ

Permissions

All chapters in this book were first published in AFE, by Springer; hereby published with permission under the Creative Commons Attribution License or equivalent. Every chapter published in this book has been scrutinized by our experts. Their significance has been extensively debated. The topics covered herein carry significant findings which will fuel the growth of the discipline. They may even be implemented as practical applications or may be referred to as a beginning point for another development.

The contributors of this book come from diverse backgrounds, making this book a truly international effort. This book will bring forth new frontiers with its revolutionizing research information and detailed analysis of the nascent developments around the world.

We would like to thank all the contributing authors for lending their expertise to make the book truly unique. They have played a crucial role in the development of this book. Without their invaluable contributions this book wouldn't have been possible. They have made vital efforts to compile up to date information on the varied aspects of this subject to make this book a valuable addition to the collection of many professionals and students.

This book was conceptualized with the vision of imparting up-to-date information and advanced data in this field. To ensure the same, a matchless editorial board was set up. Every individual on the board went through rigorous rounds of assessment to prove their worth. After which they invested a large part of their time researching and compiling the most relevant data for our readers.

The editorial board has been involved in producing this book since its inception. They have spent rigorous hours researching and exploring the diverse topics which have resulted in the successful publishing of this book. They have passed on their knowledge of decades through this book. To expedite this challenging task, the publisher supported the team at every step. A small team of assistant editors was also appointed to further simplify the editing procedure and attain best results for the readers.

Apart from the editorial board, the designing team has also invested a significant amount of their time in understanding the subject and creating the most relevant covers. They scrutinized every image to scout for the most suitable representation of the subject and create an appropriate cover for the book.

The publishing team has been an ardent support to the editorial, designing and production team. Their endless efforts to recruit the best for this project, has resulted in the accomplishment of this book. They are a veteran in the field of academics and their pool of knowledge is as vast as their experience in printing. Their expertise and guidance has proved useful at every step. Their uncompromising quality standards have made this book an exceptional effort. Their encouragement from time to time has been an inspiration for everyone.

The publisher and the editorial board hope that this book will prove to be a valuable piece of knowledge for researchers, students, practitioners and scholars across the globe.

List of Contributors

George Vachadze
Department of Political Science, Economics, and Philosophy, College of Staten Island and Graduate Center, City University of New York, 2800 Victory Blvd, Staten Island, NY 10314, USA

Zelalem G Terfa
The University of Liverpool, Liverpool, UK

Aynalem Haile
ICARDA, Aleppo, Syria

Derek Baker
ILRI, Nairobi, Kenya

Girma T Kassie
CIMMYT, Harare, Zimbabwe

Catherine Ragasa
Development Strategy and Governance Division, International Food Policy Research Institute (IFPRI), 2033 K Street, NW, Washington DC 20006, USA

Suresh C Babu
Partnership, Impact and Capacity Strengthening Unit, International Food Policy Research Institute (IFPRI), 2033 K Street, NW, Washington DC 20006, USA

John Ulimwengu
West and Central Africa Office, International Food Policy Research Institute (IFPRI), 2033 K Street, NW, Washington DC 20006, USA

Gregory Yom Din
Department of Management and Economics, the Open University of Israel, Raanana, Israel
Faculty of Exact Sciences, Tel-Aviv University, Tel-Aviv, Israel

Luigi Cembalo
Department of Agriculture, AgEcon and Policy Group, University of Naples Federico II, Via Università 96, Portici, NA 80055, Italy

Francesco Caracciolo
Department of Agriculture, AgEcon and Policy Group, University of Naples Federico II, Via Università 96, Portici, NA 80055, Italy

Giuseppina Migliore
Department of Agricultural and Forest Sciences, AgEcon and Policy Group, University of Palermo, Viale delle Scienze ed. 4, Palermo 90128, Italy

Alessia Lombardi
Department of Agriculture, AgEcon and Policy Group, University of Naples Federico II, Via Università 96, Portici, NA 80055, Italy

Giorgio Schifani
Department of Agricultural and Forest Sciences, AgEcon and Policy Group, University of Palermo, Viale delle Scienze ed. 4, Palermo 90128, Italy

Zakaria A Issahaku
Debt Management Division, Ministry of Finance, P.O. Box MB 40 Accra, Ghana

Keshav L Maharjan
Graduate School of International Development and Cooperation, Hiroshima University, 1-5-1 Kagamiyama, Higashi Hiroshima 739-8529, Japan

Maurice J Ogada
International Livestock Research Institute (ILRI), P.O. Box 30709-00100, Nairobi, Kenya

Germano Mwabu
School of Economics, University of Nairobi, P.O. Box 30179-00100, Nairobi, Kenya

Diana Muchai
School of Economics, Kenyatta University, P.O. Box 43844-00100, Nairobi, Kenya

Yasuo Ohe
Department of Food and Resource Economics, Chiba University, 648 Matsudo, Matsudo, Chiba 271-8510, Japan

Shinichi Kurihara
Department of Food and Resource Economics, Chiba University, 648 Matsudo, Matsudo, Chiba 271-8510, Japan

Shinpei Shimoura
Department of Food and Resource Economics, Chiba University, 648 Matsudo, Matsudo, Chiba 271-8510, Japan

Stefano Pascucci
Management Studies Group, Wageningen University, Hollandseweg 1, 6700EW, Wageningen (NL)

Domenico Dentoni
Management Studies Group, Wageningen University, Hollandseweg 1, 6700EW, Wageningen (NL)

Dimitrios Mitsopoulos
MSc Organic Agriculture, Consultant in Agribusiness Marketing and Innovation, Wageningen (NL)

Jeffrey Gillespie
Department of Agricultural Economics and Agribusiness, Louisiana State University Agricultural Center, Martin D Woodin Hall, Baton Rouge, LA 70803, USA

Richard Nehring
Economic Research Service, 1800 M St. NW, Washington, DC 20036, USA

Isaac Sitienei
Department of Agricultural Economics and Agribusiness, Louisiana State University, Martin D Woodin Hall, Baton Rouge, LA 70803, USA

Matthew K Loke
Department of Natural Resources and Environmental Management, University of Hawai'i at Mānoa, 1910 East-West Road, Sherman 101, Honolulu, Hawai'i 96822, USA Hawai'i Department of Agriculture, Agricultural Development Division, 1428 South King Street, Honolulu, Hawai'i 96814, USA

PingSun Leung
Department of Natural Resources and Environmental Management, University of Hawai'i at Mānoa, 1910 East-West Road, Sherman 101, Honolulu, Hawai'i 96822, USA

www.ingramcontent.com/pod-product-compliance
Lightning Source LLC
Chambersburg PA
CBHW050453200326
41458CB00014B/5170